humAIn

Library of Congress Control Number: 2024947368

ISBN Paperback: 978-1-963271-51-5
ISBN Ebook 978-1-963271-52-2

Published by Armin Lear Press, Inc.
215 W Riverside Drive, #4362
Estes Park, CO 80517

humAIn

UNLOCK YOUR POTENTIAL
USING ARTIFICIAL INTELLIGENCE

DON ROOSAN,
PharmD, PhD

ARMINLEAP

To all the bold souls who take the first step into the unknown,
knowing they might fail but still move forward with courage.

contents

COURAGE TO
EXPLORE THE UNKNOWN

If you've picked up this book, you might already have a sense of what AI and technology are all about. But even if those concepts seem like a distant galaxy, rest assured that if you've ever used a smartphone, driven a car, or tapped away on a computer, this book is absolutely for you. Whether you're an AI novice who's just stumbled upon this book out of curiosity, a tech enthusiast ready to dive deeper, or someone wary of AI's impact, this book is your companion on a journey of discovery and empowerment.

Let's start by addressing a common concern: misinformation. The big tech companies want you to be the product rather than the innovator. They flood the market with fear-mongering stories about AI, spinning doomsday scenarios and predicting a future where machines rule over humans. But here's the reality: **AI is just a tool. It's not an omnipotent force** poised to take over the world. You can control, adapt, and utilize it to improve your life.

You might be thinking, "But AI seems so complicated!" Let me tell you a secret: I've been researching AI for over a decade and still learn new things every day. Yet, I use AI daily, just like you use your phone, your car, or even a simple calculator. It's integrated into our lives in ways we often overlook. The goal of this book is to help you recognize that you're already interacting with AI and to empower you to harness its potential without fear.

AI isn't just a thing of the future—it's here, right now. It's already integrated into our daily lives in ways that seem almost mundane. Think about your virtual assistant scheduling appointments, your streaming service recommending your next favorite show, or your car's navigation system finding the fastest route home. If you don't take charge of understanding and using AI, you risk becoming just another cog in someone else's machine rather than steering your own course. In this book, we'll explore how AI impacts various aspects of life, from your professional world to your personal endeavors. We'll break down complex concepts into digestible chunks, helping you see that AI isn't a mystical, uncontrollable force but a tool you can master. And we'll debunk the myths that make AI seem like a looming threat rather than an opportunity for growth and innovation.

I recently spoke with several high school and college students who were genuinely afraid that AI would take over their future jobs. They questioned whether pursuing higher education was even worth it. I understand their concerns: AI is significantly transforming the job market. However, AI is shifting and creating new opportunities rather than eliminating jobs. The trick is under-standing where these opportunities are and preparing yourself to seize them. Not pursuing education or skills development out of fear of AI is not the solution. This concern prompted me to write this book. People like us, who are deeply involved in developing AI technologies, are extremely busy creating tools that will shape the future. Most of my colleagues don't have time to share their insights with the world because they're immersed in their work. But I felt a profound obligation to the next generation, as well as to those that have gone before. We are the stewards of knowledge, and it's our responsibility to guide those who will come after us and provide practical insights to our parents' generation. Our civilization thrives on passing wisdom from generation to generation, and that's precisely what this book aims to do.

So, what can you expect from this book? You'll gain a robust, clear understanding of what AI truly is and how you can navi-gate its complexities. We'll explore practical applications of AI, debunk common myths, and provide actionable strategies to integrate AI effectively into your life and career. Whether you're looking to boost your career prospects, enhance your daily life, or simply satisfy your curiosity, this book has something for you. Think about how AI can enhance your learning, help you stay informed, and even provide personalized support for your hobbies and interests. For instance, let's say you're a nurse worried about AI taking over your job. While AI can assist with tasks like data analysis and patient monitoring, it can't replicate compassionate care and showing human emotions. By leveraging AI to handle routine tasks, you can focus more on patient care and improving outcomes. This book will show you how to use AI as an ally rather than fearing it as a foe.

And if you're in business, you'll learn how to integrate AI to stay competitive and innovative. Companies that embrace AI are not just surviving but thriving, finding new ways to enhance customer experiences, streamline operations, and drive growth. Whether you're an entrepreneur, a manager, or an employee, understanding AI will give you a significant edge. Throughout this book, we'll also address the ethical considerations of AI. It's crucial to understand the implications of AI on privacy, security, and fairness. We'll explore how to use AI responsibly, ensuring its benefits are accessible to everyone while mitigating potential risks. I encourage you to keep an open mind as you read the chapters. Engage with the content, reflect on how it applies to your life, and consider how you can use AI to your advantage. Remember, the goal is not just to survive in the age of AI but to thrive. By understanding and embracing AI, you position yourself to lead in an ever-evolving world.

In this book, we journey through AI's dynamic and transformative landscape, designed to equip you with the knowledge and tools to thrive in an AI-driven world.

Chapter 1 sets the foundation by discussing the importance of learning, unlearning, and relearning. Our current education system teaches us to learn but often neglects the vital process of unlearning outdated methods and relearning new ones. This chapter addresses how to adapt to AI and manage the overwhelming influx of data we encounter daily. Chapter 2 demystifies AI, cutting through the hype and confusion surrounding its inner workings. We explore different types of AI and break down complex concepts into simple, understandable terms, ensuring you clearly understand what AI truly is. Chapter 3 delves into human cognition and its interaction with AI. We shed light on how we process information and make decisions by examining cognitive anchoring and other intriguing aspects of our brain's functionality. Chapter 4 examines AI's broader impact on healthcare, teams, and businesses. We look at how AI shapes the future of solopreneurs and business entrepreneurs and how these changes pave the way for innovative leadership in various industries.

Chapter 5 tackles the disruption AI is bringing to education. We discuss how this technological shift will affect future generations and the way we learn, preparing us for a future where AI plays a central role in education. In Chapter 6, we confront the issue of polarization fueled by misinformation. AI can act as a villain and a hero, helping us navigate the noise and find accurate information. This chapter provides valuable techniques for discerning truth in an age of misinformation. Chapter 7 addresses the mental health challenges exacerbated by social media and its AI algorithms. We offer insights into understanding your brain and social media habits to combat negative effects and foster healthier online behaviors. Chapter 8 focuses on digital detoxing and the crucial role of dopamine. We explain how to manage this neurotransmitter to maintain a state of calm and joy, avoiding the constant highs and lows of digital addiction. In Chapter 9, we introduce "ME thinking" – a combination of Mindfulness and Effortless thinking – to help you overcome the fear of AI. This chapter provides techniques to integrate AI into your life efficiently and confidently. Finally, Chapter 10 explores the AI revolution and its evolutionary impact. We emphasize the importance of having a higher purpose and developing a smart internalized AI. The future is not about humans versus AI but about humans becoming more humAIn. This chapter teaches you how to live a fulfilled and connected life in this new era.

So, buckle up and get ready to embark on an exciting journey. This book is your guide to demystifying AI, overcoming fear, and harnessing the power of technology to enhance your life. Whether you're a tech-savvy enthusiast, a curious skeptic, or someone just looking to stay informed, there's something here for you. Let's dive in and explore the incredible possibilities that AI offers. Together, we can shape a future where technology serves humanity, not the other way around.

part I
ADAPTATION

chapter 1

RESURGENCE WITH AI

"Unlearning is the highest form of learning." - Rumi

Once upon a time, a traveling scholar arrived at a Zen master's humble sanctuary. The scholar toyed with his favorite abacus, reflecting his esteemed abilities, but eager to gain insight into the mysteries of Zen. Confident, he believed he could impress the Zen master with easy calculation of everyday tasks.

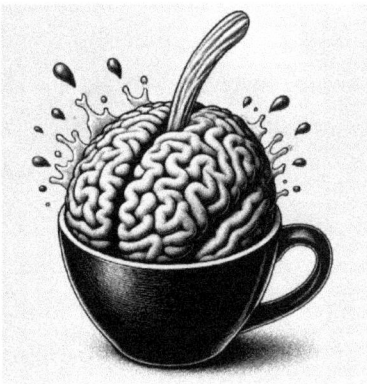

With a courteous bow, the scholar addressed the master, "I have traveled many miles to seek your wisdom. My abacus has limitations.

It can teach me arithmetic, time management, and inventory numbers. But I need more. Teach me about Zen, for I wish to understand its profound truths."

The Zen master, calm and composed, invited the scholar to sit. The subtle, sweet, earthy notes of incense mingled with the scholar's anticipation. As the scholar settled, he impatiently conveyed his thoughts and theories. Before the master spoke, the scholar shared his life and the ability to adapt to ever-changing events using his handheld abacus. The scholar talked at length about *his* interpretations of Zen, referencing ancient texts, oral histories, and learned philosophies from other practitioners from all over the countryside. The scholar persisted, staring down at his abacus instead of watching the master, ticking away at the wood beads, using the framed calculator to emphasize his learning. He asserted opinions, showcasing extensive knowledge about computing daily activities through the simplicity of the handheld device. The scholar believed the abacus was his gateway to Zen.

The master, a paragon of patience, listened to the scholar's rambling monologue, his face serene. The master offered nothing in response. The scholar continued inexhaustible tales of teaching and learning through mere calculations of his abacus; the scholar's fingers flicked through the smooth beads on the rods above and below the calculator's off-center beam. Instead of speaking, the master began preparing tea for two.

The master retrieved a teapot and two teacups. He began pouring tea into the scholar's cup. The scholar watched intently, breathing between words, wondering what wisdom this simple act might reveal. The tea flowed smoothly from the teapot to the cup to the brim. But the master did not stop. Soon, tea spilled over the edge, cascading onto the table, saturating the floor, and finally soaking into the scholar's pristine robes.

Alarmed, the scholar exclaimed, "Stop! The cup is full already. Can't you see? No more will go in!"

The master paused with a gentle smile. "Exactly," he replied. "You are like this cup – so full of your *own* ideas and opinions there is no room for anything new. How can I show you Zen unless you first empty your cup?"

The scholar sat in stunned silence. The master's words sank in. His mind, cluttered with preconceived notions, had left no space for unexplored learning and understanding. The clicking beads in the abacus had stopped. The humility of the master's simple yet profound lesson dawned upon him. The master spoke quietly, "To gain wisdom, you must first be open and receptive, shedding arrogance and making space for new insights."

As you embark on your journey through this book, remember the lesson of the overflowing cup. If you approach this information with assumptions and fears, it will be like the spilled tea, a wasted opportunity. But if you carry an empty cup, you'll be ready for wisdom.

Navigating the AI Revolution with Confidence

When ChatGPT launched at the end of 2022, it started a whirlwind of controversy and speculation. The buzz around AI has hit a fever pitch. AI stocks soared with billions of dollars flooding into AI development across various industries. Amid the excitement, waves of fearmongering and doomsday predictions abound. You've probably heard dramatic claims that AI will take over all jobs and disrupt lives irreparably. They made movies about the very nature of man vs AI technologies. Hollywood's writer's strike, highlighting AI-generated scripts and video capabilities like Google's VEO and OpenAI's SORA, adds to growing anxieties. But let's take a step back and look at AI from a more grounded perspective. We understand the ethical concerns associated with AI, and it's important to address them. AI should be developed and used responsibly, with clear guidelines and regulations in place to ensure its benefits are maximized and its potential harms are minimized.

First and foremost, AI isn't a separate entity poised to take over humanity like Skynet from *Terminator* or create unlivable *War Games* scenarios. These portrayals are pure science fiction and not reflective of reality. The same screenwriters worrying about AI taking over have planned for the worst without considering the benefits. Consider AI an incredibly powerful tool designed to assist and enhance your capabilities. Unlock the potential of AI, and you'll find a world of possibilities at your fingertips. Shifting your perspective from fear to curiosity empowers you.

Yes, AI is transforming industries, especially in tech sectors.[1]

- AI revolutionizes healthcare by assisting in early disease detection and personalized treatment plans.

- AI improves risk assessment and fraud detection.

- AI is crafting beer without working taste buds, singing karaoke, making art, and scribing sonnets.

However, this transformation doesn't mean a wholesale replacement of human jobs. Instead, AI is more likely to automate routine tasks, freeing you to focus on more complex and creative endeavors.

And how about those Hollywood writers? Sure, the idea of AI-generated scripts might seem threatening at first glance. But remember, AI lacks the nuanced understanding and emotional depth that you, as a human, bring to creative work. AI lacks unique perspectives, experiences, and emotions to make deeply resonating scripts. However, after a half-decade of lackluster box office bombs, maybe writers can use AI to shape scripts to sauce out more compelling storylines. AI can assist in brainstorming or drafting, giving writers a starting point or helping overcome writing hurdles. AI tools can churn out plot ideas at lightning speed but can't incorporate uniquely human elements worth reading. Imagine having a full-time assistant who helps generate ideas faster and builds character backgrounds, still leaving you time to make coffee and craft those perfect scenes into cohesive narratives. Don't worry about the AI beer.

Consider the tools we currently take for granted. Spellcheck and grammar software like Grammarly, Writesonic, Jasper, Copy. ai, and WordAi help you write *well*. None of those services can replace the writer. Instead, they've made you more efficient, allowing you to focus on the content rather than the writing mechanics. AI is set to play similar roles across various fields, enhancing your unique contributions through modality in technology—human-computer interaction.

So, whenever you hear a doomsday prediction about AI, think about who is telling you and their financial motivations. Are they trying to scare you because fear sells? Fear tactics have long been used to control public perception, just as education was once the domain of the wealthy few. Now, technology has democratized learning, and it's essential to embrace this change rather than fear it. By critically evaluating these fear tactics, you can empower yourself with a more balanced and informed perspective on AI and its impact, allowing you to make decisions based on facts rather than unfounded fears. Businesses and companies profit when you fear and stay behind in creating innovation that can compete or outsmart them. Next, we'll uncover the truth behind the fearmongering surrounding AI. Spoiler: It's not as scary as it seems!

The Truth Behind the Fearmongering

Right up front, you need to ignore the fear and apprehension surrounding AI, fueled by misinformation and sensationalism. Some individuals highlight potential disruptions AI could cause to the workforce. It's done to underscore the need for adaptation and preparedness. Unfortunately, this narrative often gets mis-interpreted, leading to undue anxiety. It's essential to approach AI with a balanced perspective. It's a tool designed to *enhance* and *not* replace human capabilities. Those, my friends, are what screenwriters are for.

Companies vested in promoting AI technology will change everything because it drives investment and innovation. They

might gloss over AI used as an extension of your skills and knowledge, empowering you rather than replacing you. When people predict AI taking over jobs, looking deeper is essential. AI will change the nature of many jobs, but it will also create new opportunities. Presently, AI handles data analysis, allowing professionals to focus on strategic decisions based on that data. Understanding how AI works and how to integrate it into your life is critical to staying relevant and competitive.

With AI, you can create applications without knowing complex programming languages. Thanks to advancements in large language models, AI can help you write code, design interfaces, and even troubleshoot problems, making it accessible to more people than ever before. This democratization of technology is a tremendous opportunity. However, there's a catch. Powerful companies often prefer to keep these capabilities under wraps because they threaten profit margins. Some organizations want to maintain control, making you dependent on their products and services.

Spreading fear about AI discourages you from exploring these tools on your own, keeping you as a passive consumer rather than an active creator. AI can be a game-changer for anyone willing to learn and adapt, offering a unique opportunity for personal growth and professional advancement. Whether in healthcare, finance, education, or any field using technology, AI can help you work smarter, not harder, and lead to opportunities you may not have considered before.

While it's truthful that AI will replace some jobs, it's *essential* to remember that roles demanding human creativity, empathy, and critical thinking are secure. These jobs AI cannot replicate. Preparing yourself to thrive in this evolving landscape means adapting and equipping yourself with the knowledge and insight to navigate these changes confidently.

Consider this: if your job involves significant data entry, it's wise to start learning skills complementing AI, such as project management or data analysis. In an AI-driven world, embracing lifelong learning and continuous improvement is not just a choice but necessary to stay ahead of the curve. Upskilling and reskilling are critical. Furthermore, AI can augment your existing skills. In healthcare, AI analyzes medical records and provides insight, enabling doctors to focus more on patient care. AI personalizes learning experiences in education, freeing teachers to devote more time to individual student needs. AI offers a world of opportunities for your personal and professional development.

The phenomenon of 'fear propagation' in the context of AI, spreading fear and anxiety, can often be unfounded, based on misinformation. Instead of succumbing, you should strive to understand AI and its implications. We decipher what AI is in Chapter 2 and provide you with exciting ways to learn about how to deal with the fear of AI in Chapters 8, 9, and 10. Let's consider human advancement in a different timeline than serving AI overlords. Remember, you fear what you don't understand. To understand the purpose of our information environment, you need to look deep inside the information age.

What is the Information Age?

Welcome to the Information Age, starting around the mid-twentieth century, when the quest for knowledge and connectivity constantly reshapes our world. Yes, you should care because you are in it right now! This era marks the widespread use of computers and the Internet, revolutionizing communication and information access. The Information Age is not just about having access to information—it's about how we seek, consume, and utilize technology. We don't need help from fictional alien races; we do well without outside help. The Information Age has empowered us to navigate the vast digital landscape, strategically acquiring and utilizing knowledge for our benefit.

To understand the purpose of the Information Age, let's delve into a concept called *Information Foraging Theory*.[3] This approach suggests that, like animals hunting for food, humans also "hunt" for accessible information in the vast digital landscape. In simpler terms, just as animals look for the most nutritious food with the least effort, we look for the most valuable information with the slightest exertion. This theory helps explain our behaviors in the Information Age, where efficiency and productivity are paramount in finding helpful information. The new epoch has more to do with developing our learning skills than physical interactions.

In the era of hunter-gatherers, our survival hinged on our adaptability to seek out substance in the wild, and it worked. Neanderthals weren't interested in exploring new ways to make hunting and gathering easier. They didn't want to simplify their culture. Eventually, Neanderthals would die out while humans flourished and learned to make soap, deodorant, and computers.

Similarly, you are on a constant quest for information in the Information Age. The path to success and prosperity lies in sheer effort and strategic knowledge acquisition. If you do not do it, someone else will get ahead in the life game!

The Digital Hunter

In the past, libraries, books, and experts were the primary sources of knowledge, often limited to those with access and funds.[4] Success today is about more than just hard work; it's about working intelligently and strategically using information to guide your efforts. In our digital age, being a "digital hunter" means skillfully navigating the vast landscape of online information to pinpoint where your hard work will have the most impact.[5] This involves developing effective techniques for gathering, analyzing, and utilizing data to make informed decisions that align with your goals. By honing your information-seeking skills, you can identify the most promising opportunities and focus your energy where it counts, maximizing your chances of achieving success in your endeavors. Today, search engines like Google, Bing, Duck-DuckGo, and databases like PubMed, Scopus, and ScienceDirect bring knowledge to our screens.

Online courses like Coursera and Khan Academy are available, allowing us to forage for information more efficiently than ever before.[6,7,8] It would be nearly impossible to find a specific book in a massive library without a catalog or the book title. Based on simple questions input into a search engine, thousands of relevant answers are seconds away. That's the power of information foraging in the digital age, where convenience and efficiency are at their peak, making our lives easier and more comfortable. Our tools and techniques have evolved, making our search for knowledge faster and more efficient. If only the Neanderthals had the Internet. But homo sapiens crave for instant connectivity.

Why do you desire to be connected with information all the time? Your innate curiosity and the drive to stay informed means you can make better decisions, stay updated with the latest trends, and remain competitive in your personal and professional life. Consider you're planning a trip. With constant connectivity, you can check real-time flight prices, hotel availability, and even weather conditions at your destination, enabling you to make the best travel decisions. This constant connectivity ensures you're always aware of what's happening around you, allowing you to adapt and respond quickly to changes.

The purpose of the Information Age is to harness this constant flow of learning, enhancing your productivity and decision-making capabilities. With the transformative rise of big data and analytics, you can now process vast amounts of information to uncover patterns and insights that were previously unimaginable. For many individuals, access to real-time information means making inspired choices about health, finances, education, and more. Wearable technology, like fitness trackers, provides continuous data about your physical activity and health metrics, empowering you to make better lifestyle choices. You can find lost luggage, keys, and the dog, all with a smartphone app and its algorithms on the backend.

Algorithms play a crucial role in helping you navigate the immense sea of data.[9] They are your digital foraging tools, guiding you to the most relevant and valuable information. Application of algorithms began with the invention of computers and the Turing machine, laying the foundation for modern computing. In the 1960s, algorithms made significant medical strides, helping doctors diagnose diseases and plan treatments. The use of algorithms in radiology has improved the accuracy of tumor detection. Today, sophisticated algorithms curate content tailored to your interests and needs. They help you find what you're looking for quickly and efficiently, ensuring that your foraging efforts yield the best results.

Algorithms are ubiquitous daily, curating your online experiences and tailoring content to individual preferences. They track viewing trends on social media and entertainment platforms to bring you the latest commercials in things you don't need. As technology advances, algorithms' role in medicine and beyond will only evolve, shaping a future where information is readily accessible and personalized for each individual. The ultimate goal of the Information Age is to create a symbiotic relationship between humans and technology, where technology augments your cognitive abilities and helps you stay connected while you consider context, empathy, and ethical considerations that technology lacks. Thriving in the Information Age, you must embrace continuous learning and willingness to unlearn outdated methods.

Embracing Continuous Learning and Unlearning

Rapid technological advancements mean that what you learned yesterday might be irrelevant tomorrow. Stay one step ahead of the product's commercial algorithms. One strategy for continuous learning is to attend webinars or online courses on relevant topics regularly. For adaptation, consider joining professional networks or communities where you can learn from others in the field and stay updated on industry advances.

Consider how you consume information today compared to a decade ago. Social media platforms, online news, and digital libraries transformed how we stay informed—no more flipping channels on a TV. You must continually update your strategies and tools to remain an effective forager. The Information Age is not just a dynamic period but an exciting one that offers unprecedented opportunities for growth and innovation.[10] Understanding and embracing AI ensures you always have access to the information you need and inspires you to seize this era's potential. In the next chapter, we demystify AI and you will feel more confident. Knowledge always empowers.

Your role is to stay curious and adaptable, leveraging technology's power to enhance your decision-making. Embrace the symbiotic relationship with technology and use it to amplify your capabilities and productivity. Remember, the goal is to survive and thrive in this ever-evolving landscape. It's crucial to unlearn old ways and let go of outdated methods in this era of rapid change, which is a sign of intelligence.

IN73LL1G3NCE IS 4HE AB1L17Y T0 AD4PT T0 CH4NG3

If you can read the above, don't be alarmed cause your brain is learning to unlearn and relearn in an instant here.

Unlearning is vital to your personal development as it shapes your ability to change.[11] Unlearning doesn't mean discarding everything you know. It's about embracing new opportunities and recognizing when certain methods or beliefs no longer serve your quest. Unlearning is the process of letting go of outdated or incorrect knowledge, habits, or assumptions to make room for new understanding and skills. It involves critically examining what you know and questioning long-held beliefs that may no longer serve you or align with current realities. Unlike relearning, which involves refreshing or updating existing knowledge, unlearning requires a conscious effort to dismantle entrenched patterns and embrace change. This can be challenging because it involves overcoming cognitive biases and emotional attachments to familiar ideas. However, unlearning is essential for growth and adaptation in a rapidly changing world, enabling individuals to remain flexible and open to new possibilities, perspectives, and innovations. By unlearning, you create space for continuous learning and development, allowing you to thrive in dynamic environments. By unlearning, you can elevate your existing platform. Holding onto these obsolete practices can hinder your progress.

Welcoming new ideas allows you to stay ahead of the curve and propel your business forward. Remember, while it's important to adapt to new technologies, the fundamentals of reading, writing, and arithmetic are still crucial.

In the current digital world, adaptability is a necessity.[12] Failing to innovate and evolve with changing market dynamics often leaves companies outmoded by agile competitors. This underscores the importance of strong leadership encouraging the unlearning process. Consider Eastman Kodak, once a photography industry giant. Despite being a pioneer in technology, Kodak's reluctance to embrace digital photography led to its downfall.

On the other hand, Canon and Nikon, adapting to digital trends, continue to thrive. Similarly, traditional brick-and-mortar stores resisting the transition to e-commerce face significant challenges in retail. However, companies embracing online sales and digital customer engagement strategies have experienced sustained growth. The success stories of Amazon and Shopify underscore the importance of unlearning old business models and adopting new ones. Today, instead of going to a grocery store, you can use your smartphone to have an associate navigate the store aisles and shop. You don't have to catch up with neighbors, run into old friends, or carry groceries to your vehicle. Instead, you enjoy the comfort of your home and use your phone for other tasks.

Clinging to outdated beliefs hinders your progress, preventing you from realizing your full potential. For instance, if you were raised to fear AI overlords, you might refrain from taking risks and exploring ways to overcome digital threats. A growth mindset—seeing risk as a chance to learn and grow—is not just beneficial; it's essential. Consider how your approach to learning has transformed over time. Education might have been more about memorization and standardized testing in the past. Today, it's all about critical thinking, problem-solving, and adaptability. To help you in this process, technology is your best friend.

Technology plays a significant role in the job market, constantly changing with the rapid development of artificial intelligence, machine learning, and automation.[13] To stay relevant, online learning platforms, coding boot camps, and professional development courses are not just resources. They are your allies who keep you ahead of the curve. These platforms offer various courses and programs. They provide a supportive learning environment where you can interact with instructors and fellow learners and get feedback on your progress. Embrace these opportunities to ensure you're prepared for the ever-changing job market. Unlearning requires a growth mindset—believing your abilities and intelligence can develop through dedication and hard work. Those saber-tooth kitties are still lurking in the high grass plains, but they're digital copies of their former selves, and you're a master hunter in the digital age. This mindset encourages you to seek new challenges, learn from your experiences, and continually strive to improve.

In the Information Age, a growth mindset means being curious and excited about new technologies. Your willingness to experiment and take risks, knowing that even if you fail, you'll learn valuable lessons, will help you succeed. In the age of massive data points captured daily, you must learn how to adapt to take advantage rather than become a product for technology companies. Let's look into how the data explosion in recent years is shaping the environment around us.

Surviving Data Deluge

In the twentieth century, news consumption was a straightforward process. People relied on newspapers or TV, trusted sources that diligently ensured accuracy and consistency. However, with the advent of the twenty-first century and the Internet, this simplicity was replaced by a complex landscape of information and capitalizing fearmongers.

Now, with a few clicks, anyone, anywhere, can share information online. It's challenging to tell if what we're reading is true because fact-checking is an entirely different set of algorithms. It used to be information was disseminated and vetted by scholars and professionals. Nowadays, anyone in the corner office has a new app that creates fake media only to entertain himself and cause distrust in the media. The tsunami of misinformation and disinformation is higher than any beachfront property on the planet.

Mark Twain once said: "A lie is halfway around the world before the truth has got its boots on." And if you're falling down the rabbit hole of misinformation, you'd believe the quote came from Churchill and not Twain. But you're mistaken on both accounts. The poet Vigil coined the phrase in BCE, and because Twain and Churchill could use what they had unlearned, you'd believe both of them had come up with the quote.

Every single day, our world generates over 2.5 quintillion bytes of data. That's a staggering number with eighteen zeros! And it's not showing any signs of slowing down. In 2020, we amassed about forty zettabytes of data. That's equivalent to forty trillion gigabytes. But hold your breath because, by 2025, it's projected to reach a mind-boggling 175 zettabytes. The proliferation of social media, internet-connected devices, online shopping, and cloud computing for data storage is unstoppable. And does anyone know if all that additional digital data has a tangible weight?

That tsunami of data is staggering and surfable. On one side, it's a game-changer, fueling advancements in AI and personalized marketing. But on the flip side, it's a tightrope we grapple with safeguarding data and respecting individuals' privacy. Consider the medical field. It's inundated with data, doubling every 73 days. The deluge of electronic health records and medical images can empower doctors to provide highly personalized care and make groundbreaking discoveries. However, it poses a significant hurdle as healthcare professionals must continually learn and process the right technology to make sense of diagnoses.

Conquering this data tsunami, we need intelligent data management, serious privacy software, and teamwork across different areas of expertise. AI is becoming our go-to tool for understanding, cataloging, and using this massive data wave. The data deluge affects everyone regardless of the industry—healthcare, finance, entertainment, or logistics. Technology makes it super easy for humans to handle individually. This gives you unique opportunities to learn and improve what you do. It means you've got to figure out how to surf, glide, and use all this data in the best way possible to achieve your point break. Remember, you fear what you don't understand. In Chapter 2, we explained AI in very simple terms so you can eliminate the fear once and for all!

First, we shape our tools, thereafter they shape us

Have you ever noticed how AI has seamlessly woven itself into the fabric of our daily lives? From the moment you wake up to your virtual assistant's cheerful greeting to the late-night Netflix recommendations, social media posts, and latest trends in making omelets. AI is everywhere, making life a bit easier and much more exciting. AI mimics human behavior without distractions, caffeine, and bathroom breaks. These machines use complex algorithms to process enormous amounts of data every millisecond. They recognize images, interpret signals, and make real-time

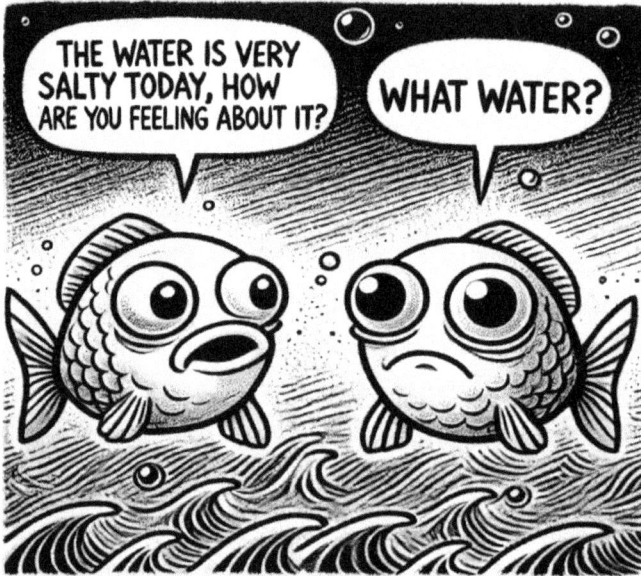

decisions to navigate hazards safely. Imagine your car calculating and recalculating the optimal path, adjusting to changing environments, and ensuring your safety—all without you lifting a finger or your foot from the accelerator pedal. It's like having a super-intelligent co-pilot who never gets tired and can keep a lookout when you're distracted, low on caffeine, and looking for a restroom during your travel.

AI's influence extends far beyond our roads. In pharmaceuticals, AI sifts through vast datasets to identify potential compounds faster and more accurately than ever. This accelerates the development of new treatments and brings us closer to the era of personalized medicine, where treatments are tailored specifically to individuals. In social media, AI operates in the background, analyzing your behavior, curating content, and predicting trends, ensuring those product commercials are customized to your viewing trends. It's why your feeds always align with your interests, keeping you engaged and connected. AI-powered algorithms monitor transactions for fraud, bolstering the security of your finances. AI elevates customer experiences in retail with person-alized recommendations, dynamic pricing, and efficient inventory management. The essence of AI lies in deciphering the meaning of our reality and the vast data surrounding us.

You are constantly generating data from when you wake up to when you go to sleep. Every interaction on social media, every digital footprint on various platforms, every click, and every "like" contributes to an ever-growing wave of information. This con-stant data stream reflects our digital lives and efforts. It is nearly impossible for the human mind to sift through this data flood without help. AI steps in as a practical solution to manage and utilize data so you won't drown under the tsunami.

Consider your daily interactions with search engines like Google. When you type a query, Google indexes millions of web pages to present the most relevant results. Traditionally, this process required you to sift through multiple links, skimming through information until you found what you needed. With advancements like ChatGPT and Google Gemini, your search experience is not just improved, it's transformed. These AI tools offer a more refined understanding of your queries, providing tai-lored responses. Instead of browsing through endless links, you get precise, relevant information, making your online research more efficient and effective.

Let's say you're searching for some health information. In the past, you might have had to read through dozens of articles to gain a comprehensive understanding. Today, AI can aggregate data from reputable sources, summarize it, and present a concise, accurate overview. This boosts your efficiency and deepens your understanding, empowering you to make informed decisions about your health. The AI's ability to filter irrelevant information and highlight what matters most is revolutionary. This remarkable capacity of AI also presents challenges. The same technology enhancing your search results can also manipulate the information you receive. Companies pay search engines to prioritize certain content, potentially distorting the data presented to you. This creates the weight of the tsunami: while AI can streamline your access to information, it can also restrict it, presenting only what certain entities want you to see. Some of that product advertising in your social media feeds is not always the best solution for you, especially medicines.

This underscores utilizing AI to enhance your life effectively. The notion AI is a mysterious, all-powerful force that will change everything is misleading. AI is a tool, and like any tool, its value depends on how you use it. To harness its full potential, you must treat it like an instrument, not a weapon. By doing so, you can master AI. But to truly benefit from advancements, you must be willing to unlearn and adopt new habits. The rapid pace of technological change means that clinging to outdated methods can hinder your progress.

Similarly, holding on to outdated beliefs in your personal development can prevent you from growing and adapting to new realities. It's not just about adapting to AI. It's about using it to your advantage in the Information Age. You can truly utilize AI through unlearning and relearning the digital landscape.

As you navigate the Information Age, remember that continuous learning and adaptability are your most powerful tools.

Embrace new technologies, stay curious, and be willing to bring the empty teacup. This journey is not just about surviving the Information Age; it's about thriving in it, using the wealth of information at your fingertips to create a better, more connected world. Understanding AI as a tool, not some distant object, is the first step on this inspiring journey to become the next productive humAIn.

In the next chapter, we will demystify AI. Like the empty teacup we discussed earlier, be the scholar who keeps his head up, watching for the changes. Approach this with an open mind, ready to absorb new insights. You'll discover that AI is a tool designed to make your life easier, not something to be feared.

Challenge your thinking

- Whenever you encounter a doomsday prediction about AI, consider the source and their potential financial motivations. Are they affiliated with companies profiting from fear to drive their own interests? Doing so empowers you to think critically.

- Historically, fear tactics have been practical tools for the wealthy to control the less privileged. Access to new technologies is now manipulated through fearmongering to deter broader understanding and adoption. Does this perspective align with your observations?

chapter 2

Hello AI

If you can't change the direction of the wind, adjust your sails to always reach your destination. - Jimmy Dean

The Pioneers and the Printing Press

Remember the 1400s, enjoying a quiet evening at the tavern? Suddenly, a friend you haven't seen in a while shows up and sits down to exclaim, "Have you heard about Gutenberg's new machine?"

Back then, books were rare treasures, painstakingly copied by monks. Knowledge was a luxury reserved for the elite. Much of the general population didn't know how to read. Johannes Gutenberg, a German metal worker and inventor, had a revolutionary idea to mass-produce written words. The scribes were horrified, fearing it would ruin their craft. Despite their protests, Gutenberg pressed on.

In 1455, the printing press upended the world. People had to unlearn what they had feared and learn a better way to gain wisdom. Books, once the realm of the privileged, transformed learning, igniting the Renaissance and the Reformation and disseminating knowledge across the flat earth. Conversely, much like today, booksellers were chased out of villages; fear and ignorance caused the destruction of some printing presses. Today's AI echoes this revolution, transforming society by democratizing information.

This narrative is a stark reminder of the weighty responsibility accompanying technological advancements. Today, AI is our modern-day printing press, with the potential to reshape our world by automating tasks, fostering creativity, and making information accessible. However, like the printing press, AI must be developed carefully to ensure fairness, transparency, and accountability. The consequences of mishandling AI could be far-reaching, underscoring the importance of our collective vigilance in its development. Gutenberg died a poor man because he had partnered with lawyers to create the printing press. His partners didn't strive to improve it; they just took it away from the inventor. AI's future, seized by big corporations, can feel like a struggle.

Delving into the basics of AI and the concept of the "black box," each of us has a crucial role in this narrative. You can actively engage and advocate for responsible AI in all aspects of your life. AI's inherent internal workings or decision-making processes are mysterious to humans. You give AI data and it spits out conclusions and decisions that make sense. Fearing the printing press created misdirection. Many considered it a disastrous blot on their society. AI can do more than fill in missing words. Unlike the printing press, much of its capabilities are still inside the so-called black box.[2] Much like in the past, when printing was considered, many copycats stole the idea after Gutenberg's press, creating other versions of the printing press. The same could be said about AI.

We'll dive into the mysterious black box later in this chapter.

Let's discuss how AI is integrated into your daily lives so you better understand its inner workings.

We Become Tools Of Our Tools

Once upon a time, you traveled to get to a phone. Now phones travel with us. The portable AI genius means you're never truly alone. That voice assistant who answers your questions, sets reminders, gets you lost down closed roads that haven't been updated by Google Maps, and tells you jokes? AI. And the facial recognition software that unlocks your phone or locks you out because you just woke up still wearing your clown makeup from the masquerade party last night and enhanced your security, that's also AI. The predictive text finishes your words before you can think of them, adding phrases you didn't want to use, making your typing faster and more accurate. You guessed it – AI.

Remember those nerve-wracking parallel parking maneuvers? The only reason you failed your driving test *three* times? At the push of a button, your car can park itself. Self-driving cars powered by cutting-edge technology to handle the operation and maneuvering are not just a futuristic dream; they're a reality tested worldwide.[14] Tesla's Autopilot system is another example, demonstrating AI's life-saving potential with advanced driver-assistance features. Waymo's cars are already navigating streets and picking up passengers in Phoenix, Arizona, without human intervention.[15] Black box technology works without bathroom breaks and distractions. The concept of humans riding in the back seat of driverless cars is as frightening to the wealthy as printed books were when Guttenberg changed the game. AI-powered cars don't just mean convenience—they also promise enhanced safety. According to the National Highway Traffic Safety Administration, 94 percent of serious crashes are due to human error.[16] AI, on the other hand, doesn't get distracted or tired, significantly reducing this risk. It's not just theoretical promises but reality demonstrated by Nuro's driverless delivery vehicles in Houston, Texas, which have already improved mobility for the elderly and disabled. Embrace the ride and ignore the empty driver's seat.

From diagnosing diseases to developing personalized treatment plans, AI is transforming the medical field. Machine learning algorithms can analyze medical images like X-rays and CT scans with remarkable accuracy, often spotting details that even experienced doctors might overlook. AI-powered chatbots provide mental health support and answer some medical questions. And AI is even being used to discover new drugs and treatments.[17] While AI won't replace doctors anytime soon, it's rapidly becoming their trusted and indispensable sidekick, enhancing their capabilities and improving patient care. The most coveted field in medicine once was radiology, where radiologists used to make $600,000 on average yearly. Now, primarily, AI is used to detect anomalies from images with close to perfect accuracy. Nonetheless, those radiologists still make $600,000, but the AI assistants can disseminate data faster than the doctors can read the results. We'll need those radiologists to decipher the data for the general public.

In future chapters, we'll discuss AI's use and abuse on social media. First, let's examine how AI actually works and what AI is.

But What Is AI and How Can It Do My Work?

Artificial Intelligence, at its core, is a fascinating teaching machine that replicates human intelligence, learning, reasoning, and problem-solving. It perceives its surroundings and understands languages.[18] It does not just harness raw computing power to take over the world; it gives machines a semblance of cognitive abilities, making them almost human-like in their capabilities. So, taking over the world is easier.

Three key players form the backbone of AI:

- Machine Learning (ML) is the foundation of AI. ML involves training computers to learn from data, identify patterns, and make predictions without explicit instructions. Picture ML as a computer's way of learning through experience, much like how you learn to ride a bike by practicing.

- Deep Learning (DL) is a subset of ML intricately woven network like the human brain's neural pathways. DL performs challenging and exciting tasks, deciphering images and comprehending natural languages. Picture DL like the brain within the computer, tirelessly crunching vast amounts of data to unravel the world's complexities without losing sleep or needing caffeine.

- Natural Language Processing (NLP) is the
remarkable link between humans and machines.
NLP enables computers to understand, interpret,
and generate human language in a manner that's
becoming almost indistinguishable from our own.
It's like conversing with a friend who happens to
be a computer, bridging the gap between human
communication and machine understanding. Large
language models such as ChatGPT, Gemini, or
Claude use advanced NLP and transformer models,
an ML type, to understand human language
accurately. All of this happens inside the black box.

AI has revolutionized tasks that were once the exclusive
domain of human intelligence. For instance, the voiceover nar-
rates those product commercials popping up on your media feeds.
And AI's efficiency and accuracy are not stagnant but constantly
improving. However, capturing the headlines today is the deep
learning algorithms, showcasing their efficiency in our daily lives.

A Deep Dive into Deep Learning
Deep learning is like a vast and intricate web of neurons within
a computer. Each of these neurons is like a tiny cubical worker,
processing bits of information, passing signals to one another, and
constantly adjusting connections based on the data received.[19]
This dynamic training is the heart of deep learning. The network
recognizes patterns, making increasingly accurate predictions over
time. Imagine a spider's web. Each node in the web represents

a neuron, and the threads connecting them are the pathways through which information flows. If a digital insect gets caught in the web, snapping the threads, the spider can wrap it up and send it back to the cubical workers to learn from the flaws, making a stronger, more reliable web. Introducing new data, the web adjusts itself, strengthening connections and weakening others, refining its structure to better understand the patterns within the data. This adaptability is what makes deep learning so impressive and so relatable.

Deep learning is a revolutionary method enabling computers to learn and make decisions independently, discovering from experiences. Computers use mathematics and patterns to deduce information. Think about teaching a computer to identify different animals in pictures. By exposing AI to multiple images of cats, dogs, and birds and providing markers for every animal, deep learning allows the computer to automatically identify patterns in pictures distinguishing one animal from another. This is accomplished by creating numerous layers of so-called detectors. The first layer of detectives might only see edges or corners. The next layer might build those combinations of edges to form shapes, like a cat's ear or a dog's nose. As the investigation deepens, detectives build on complex clues, recreating the whole face of a cat or the body of a dog. Over time, the last detective lays out the clues; the team can confidently say whether it's a cat, a dog, or something else.

Deep learning distinguishes between cats, dogs, or an egg sandwich by layering markers to make up a mosaic that eventually creates the image. These signals are snippets of information and have different strengths or "weights." The classification adds up the weighted patterns it receives. If the sum is strong enough, it fills in the gaps. Pieces of the mosaic are minute but enough to build on, making sense of the image. These networks of interconnected layers make decisions based on prediction models from historical evidence. Deep learning models have many layers building on simple features, like edges in an image or basic sounds in speech. As information moves through the layers, it becomes more complex. Later layers might recognize shapes, objects, or even entire words and sentences. Once assembled, the mosaic creates the finished product.

One of the most awe-inspiring aspects of deep learning is its autonomy. Deep learning systems, or detectives, aren't created by humans. They learn to become detectives on their own by studying numerous examples. This makes deep learning incredibly powerful because it uncovers patterns humans overlook. This autonomy is a testament to the potential of deep learning and its ability to surpass human capabilities.

Deep learning technology empowers many things you interact with every day. It's the reason search engines can find what you're looking for with a few hints. Deep learning gives online stores suggestions for products you might like and how self-driving cars understand the difference between roads and sidewalks. Deep learning works tirelessly behind the scenes, using those cubicle workers to strengthen the web, making sense of the vast data and improving your interactions with technology. Deep learning makes everything seamless and efficient, making the subtle distinctions between the cat, dog, and egg sandwich.

Despite the amazing intelligence that arises through deep learning that we understand, we could not figure out some part of the black box of "HOW" embedded in the emergence of deep learning.

Black Box, The Cat That Was or Wasn't There

Deep learning is a powerful tool revolutionizing various fields. However, it also presents a challenge because we don't know what's happening inside the black box. And it has significant real-world implications. Remember that pesky cat inside Erwin Schrödinger's box?[20] Is it there, or is it swimming through the vast infinity? The black box goes beyond the physics of an existing or nonexistent feline. Even experts find it challenging to fully grasp the inner workings of these models. It's as if the network's intelligence emerges from the complex interplay of data and algorithms, making it harder to pinpoint the exact mechanisms behind its decisions.

The black-box problem, like trying to understand a magician's trick, is that you see the start and the end of the trick, but the middle—the actual process of how the trick is performed—remains a mystery. In the context of deep learning, you can see the input data (like images, text, or sounds) and the output (the classification, prediction, or decision made by the AI), but the process by which the AI arrives at that output is not always clear. How did Schrödinger make that cat in the box disappear?

Let's return to AI, knowing the difference between cats, dogs, and sandwiches. You feed the model a picture, and it accurately tells you whether it's a cat, dog, or egg-on-rye sandwich. However, if you were to ask the model to explain why it made that decision, it can't define its reasoning in an easily understandable way. The neural network has learned from thousands of examples, constantly refining the connections. These adjustments are not straightforward rules but rather the vast webs of cubical workers strengthening the accurate outcome. AI demonstrates intelligence by recognizing patterns, making predictions, and even generating creative content. For example, when GPT models generate coherent text, it's not because they "understand" language like humans do. Instead, they have processed vast amounts of data to identify patterns and use those patterns to produce outputs that seem intelligent. This emergent behavior results from the collective functioning of the neural network's many interconnected nodes. However, the exact mechanisms remain a mystery due to the sheer complexity of these systems.

One field that has had a transformative impact on the black box problem is radiology. Deep learning algorithms now analyze medical images with remarkable accuracy, such as X-rays or CT or MRI scans. These algorithms can detect abnormalities based on natural bodily imaging and diagnose at a level that rivals—and sometimes surpasses—human radiologists. This transformation isn't about replacing doctors but augmenting their capabilities. By providing a sharper understanding of the mosaics making up the human body, these algorithms can ensure more accurate diagnoses, reducing the chances of human error.

The opaque nature of black-box deep learning empowers AI to perform incredibly complex tasks that go infinity and yonder exclusive domain of human intelligence. However, it also necessitates the development of better methods for interpreting and understanding these models. This is not just important but crucial for ensuring trust and transparency in AI systems, particularly in critical areas like healthcare. The more we understand how AI arrives at its decisions, the more we can trust its outcomes, and the more ethically we can use it. We'll eventually understand the difference between Schrödinger's kitty in the box and that egg sandwich.

One area of concern for the black box that concerns all of us is, of course, social media algorithms.

Silent AI Ninja in Your Social Media

Ever found yourself doomscrolling or doomsurfing through your social media feed, wondering why certain posts appear just when you need them? That's no coincidence; it's AI at work. These sophisticated AI algorithms meticulously analyze your behavior, preferences, and interactions to craft a personalized stream of content tailored just for you. The aim? To keep you engrossed, skimming, and clicking; this engagement is how social media platforms generate revenue.

These algorithms anticipate your online habits, identifying what you like, what you agree with, and even what irritates you. They are not merely about showing you tailored ads; they aim to feed your biases. By learning your preferences, they create echo chambers, where you're primarily exposed to content that reinforces your existing beliefs.[21] This phenomenon significantly contributes to the polarization of public opinion and the dangerous spread of misinformation, which can have serious societal implications. We discuss this in depth in Chapter 6: Cutting through misinformation in a polarized world.

Of course, there are always challenges with evolving technologies. Internet Addiction (IA) exemplifies one such challenge, as it involves the overstimulation of pleasure receptors in some individuals.[22] This issue is particularly relevant when considering certain AI-related factors. Currently, the brain's reward circuit function, decision-making ability, and executive function of individuals with IA are significantly lower than those of healthy individuals.

Not every form of addiction goes into your body through your mouth, nose or other places. Sometimes, it's as simple as hand to eye to brain. The pleasure you derive from dopamine hits compels you to seek more, creating a feedback loop that's hard to break. Social media platforms capitalize on constantly refining their algorithms to better understand and cater to your preferences, ensuring a steady stream of dopamine-releasing content. We discuss the effects of dopamine and how one can effectively deal with dopamine regulation in Chapter 8: Digital Detox to Reclaim Your Brain from dopamine overload.

We discuss the implications of powerful social media algorithms and how you can be ahead of the curve by utilizing some life-changing skills in Chapter 7: The Hidden Impact of Mental Health in Social Media's AI Algorithms. It highlights the importance of being mindful of your media consumption and seeking out diverse sources of information to break free from the algorithm-driven echo chambers. This is a responsibility we all share, especially for the next generation.

The Parents' Nightmares

Parents have handed off babysitting to AI long before anyone realized social media was a carefully crafted problem. Social media platforms use AI to curate content based on user ages and preferences, which can inadvertently reinforce existing biases. This particularly impacts children and teenagers, whose views and beliefs are still forming. Suppose a teenager frequently interacts with content about a particular hobby like javelin throwing or controversial topics like how lawn darts are just tiny javelins and easier to throw at other people. In that case, the algorithm will feed them more of the same, potentially narrowing their perspective. Old-fashioned lawn darts were *always* dangerous and worked better than javelins because they were easier to carry around than a light spear designed as a ranged throwing weapon. No one questioned lawn darts on the school bus, but everyone noticed a javelin. Years ago, some companies thought making candy cigarettes was a good idea. Suddenly, it's frowned upon when you get kids addicted to something everyone thought was benign.

According to the CDC, suicide is the second leading cause of death among individuals aged 10-34 in the United States. Additionally, the National Institute of Mental Health (NIMH) reports approximately 3.2 million adolescents aged 12-17 in the U.S. had at least one major depressive episode in 2019. In China, with an internet penetration rate of over 94.9 percent, the prevalence of IA is surging. In particular, the proportion of underage internet users with internet dependency psychology has been as high as 17.3 percent. This condition can lead to social withdrawal tendencies and suicidal tendencies.[23]

Anyone frequently interacting with content about a particular hobby or controversial topic will fall down digital rabbit holes. The algorithm will feed them more potentially narrowing perspectives. Over time, this can create an echo chamber, limiting your exposure to diverse viewpoints and reinforcing a skewed worldview. This is how the Flat Earth Society still exists. You must actively monitor and understand what your children are watching. This isn't about spying but ensuring that the AI algorithms broaden your child's understanding rather than reinforcing narrow viewpoints.

As a parent, you have the power to influence how your children interact with social media. By understanding the biases of AI algorithms and taking proactive steps to monitor and guide your children's online activities, you can help prevent the negative impacts of social media on their mental health. Stay vigilant, stay engaged, and help your children navigate the digital world safely. Your involvement could make all the difference in their mental well-being and overall development. Don't be afraid to talk about controversial topics and engage in seeking the right answer to questions. Understanding how AI shapes our children's brains and cognition is vital, as discussed in the next chapter.

The Ethical Tightrope

AI carries a host of ethical considerations. How do you ensure data privacy when AI systems constantly collect and analyze vast amounts of personal information? What about algorithmic bias, where AI systems inadvertently discriminate against certain groups based on race, gender, or other factors? As AI becomes increasingly capable, what does it mean for the future of human employment?

These are questions that society is grappling with, and the answers are ever-changing. However, one thing is sure: the ethical implications of AI are too important to ignore. You must ensure that AI is developed and used responsibly, with transparency, accountability, and a commitment to fairness and equity. One of the crucial issues that can dramatically affect the effectiveness and fairness of AI in healthcare is bias in training datasets. Let's go back to healthcare AI algorithms trained on data that under-represents specific populations. The AI can produce biased and unjust outcomes not out of spite but out of mishandling input. These biases often reflect broader societal inequalities related to socioeconomic status, race, ethnic background, religion, gender, disability, or sexual orientation.

Take, for example, an interesting fact about one of the multibillion-dollar American healthcare technology-based firms that had specialized AI systems at work behind the scenes in thousands of doctor offices. People praised the system for its plug-and-play aspects, diagnosing and medicating millions of patients. But, upon closer look into the data revenue, one thing became abundantly clear. The system worked, but it was flawed. You see, African American patients had higher blood pressure, and their diabetes was more acute. In its assessments of risk, the algorithm was perpetuating bias. Because African American people spent less on health care, the algorithm had learned to recommend that individual African American patients be given half the amount of care as white patients.[24]

Addressing these biases is necessary and urgent to ensure that AI-based decision support systems in healthcare are fair and accurate. So, diversifying training datasets is essential. You can help AI systems better understand and serve diverse populations by including a wide array of demographics in datasets. First, collecting data from various sources to capture the full spectrum of human diversity, ensuring that no group is underrepresented, is essential. Second, implementing fairness criteria in algorithm design is critical. Creating specific goals and metrics for fairness while developing AI systems should never stop. By incorporating fairness checks at every stage—from data collection to model training and deployment—you can create AI systems that are more equitable. Finally, continuous monitoring of outcomes is vital. Even after deployment, AI systems must be regularly assessed for performance across different demographic groups. This ongoing evaluation helps detect and correct potential biases, ensuring that the AI system remains fair and effective.

United We Stand, Divided We Fall

Thankfully, awareness of AI bias, which refers to systematic and repeatable errors in systems that create unfair outcomes is growing. Thanks to due diligence, facial recognition software can distinguish between your face, the cat, the dog, and an egg sandwich. Governments and companies are taking steps to constantly address AI biases. Many organizations now prioritize diversity and inclusion in their AI teams, ensuring that different perspectives are represented in the development process. They're also developing tools and techniques to detect and mitigate algorithm bias.

Ensuring fairness and accuracy in AI for healthcare is crucial to avoid issues like misdiagnosis. Fundamental steps toward this goal are using diverse and representative data to train these AI systems. AI can be just like humans—full of quirks, biases, and occasional bad habits. Teaching these algorithms using data from our imperfect world, with all its juicy drama around race, gender, and social status, results in interesting outcomes. Once these algorithms get rolling, they become enigmatic wizards in black boxes, leaving only the tech-savvy sorcerers among us to figure out their magical mumbo jumbo.

Clinical experts are the backbone of this process. By involving pharmacists, doctors, and other healthcare professionals in developing AI systems, we can ensure these technologies align with real-world clinical practices. These experts provide invaluable insights beyond what raw data can offer, fine-tuning AI to be practical and reliable in everyday healthcare scenarios. Their involvement is reassuring, bridging the gap between theoretical AI models and practical applications and making the systems more relevant and trustworthy. Real-world applications for strong AI arguments come from the surge in advancements in medical diagnoses using AI. Evaluating the constant input allows for a broader picture in the medical field and real-time answers.

Transparency in AI decision-making is of utmost importance. When an AI system is used to suggest diagnoses or treatments, understanding how these recommendations are made is crucial. Clear and transparent decision-making processes build trust among healthcare providers and patients, empowering doctors to confidently explain AI-driven recommendations. This transparency not only ensures accountability but also fosters a sense of confidence in the AI system. This is the reason that a radiologist will never be obsolete.

From the industry side, Google has introduced a tool called the What-If Tool. This tool allows developers to test how their AI models perform on different subgroups of data, helping them identify potential biases. It is a free online interface: You can test performance in hypothetical situations, analyze the importance of different data features, and visualize model behavior across multiple models and subsets of input data, and for different ML fairness metrics.[25] Similarly, IBM has launched the AI Fairness 360 toolkit. This toolkit provides a comprehensive set of algorithms and metrics for measuring and mitigating bias in AI systems. It includes tools for checking bias in data, bias in models, and bias in outcomes, ensuring a thorough approach to addressing AI bias.

Governments are also getting involved. The European Union's General Data Protection Regulation (GDPR) includes provisions aimed at protecting individuals from discriminatory decisions made by automated systems, including AI. These provisions require organizations to provide explanations for AI-driven decisions, giving individuals the right to challenge these decisions. In the United States, there are ongoing discussions about the need for AI regulations to ensure fairness and accountability. These discussions are considering various approaches, including establishing regulatory bodies and developing industry standards for AI. Many lawyers are involved in making sure everyone gets rich from the right government plan.

Why It Matters to You

You might be thinking, "I'm just an average person, not some tech guru or politician. Why should I care about AI bias?"

AI isn't some distant, abstract concept confined to a research lab; it is a real-world application constantly in use constantly evolving. Algorithms determine the job listings you see on your LinkedIn, Indeed or ZipRecruiter. AI evaluates your creditworthiness when you seek a loan. The software influences the social media posts you encounter and even the neighborhoods that receive more police attention.

Suppose you're a highly qualified woman applying for a tech job. Unbeknownst to you, the company's AI-powered screening tool has been trained on historical data that predominantly features male candidates. This subtle bias might cause the system to rank your application lower than it deserves, potentially leading to a missed opportunity despite your perfect fit for the role. This is an opportunity lost, not due to your skills, but because of a hidden prejudice ingrained in the AI. You can take courses in drafting the perfect AI-friendly resume to ensure yours is at the top of the heap when HR opens their emails. You could lose to the tech nerd sitting in the corner office rising high because he knows how to manipulate an outdated AI algorithm.

You have a solid credit history, but the bank's AI-powered loan approval system has learned from past data that people from your zip code are statistically more likely to default. As a result, you might be denied the loan or offered it at a higher interest rate despite your solid financial standing. That directly impacts your financial well-being, not based on your merit but on a biased generalization.

AI bias has real-world consequences for people's lives, livelihoods, and opportunities. It's the modern-day equivalent of redlining, a practice where discriminatory practices are embedded in seemingly neutral systems, perpetuating inequality and injustice. Just as redlining—a discriminatory practice consisting of systematic denial of services such as mortgages, insurance, and other financial services to residents of certain areas based on race or ethnicity—was a stain on our history, AI bias is a stain on our present, and we must address it.

The Only Thing Necessary for The Triumph Of Evil Is For Good People To Do Nothing

We can't just sit back and hope for the best when AI takes over the world. We need to be informed, vigilant, and proactive. By demanding transparency from companies using AI, we can shine a light on biased algorithms and hold them accountable. Advocating for regulations prioritizing fairness and equity is another crucial step. Governments need to create a level playing field for everyone. The GDPR is a legislative body that protects individuals from unfair automated decisions, including AI. In the U.S., discussions are ongoing about similar regulations, such as the Algorithmic Accountability Act and the Ethical Use of Artificial Intelligence Act. By staying informed and participating in public consultations, you can help push these necessary changes forward.

Addressing AI bias is undoubtedly a complex task. The ease with which biased AI can be created, even by individuals in their garages or basements, without considering its broader impact, poses significant challenges. This ease of development makes it difficult to maintain oversight and enforce standards consistently, particularly within large companies. The lack of a practical, comprehensive oversight of all AI projects further complicates matters. However, your voice holds immense power. We must collectively demand transparency from those who develop and deploy AI.

Supporting organizations dedicated to developing ethical AI can also make a significant impact. The Partnership on AI (PAI) brings together researchers, academics, and industry leaders to develop best practices for AI. They are a global multi-stakeholder nonprofit addressing the social implications of AI technologies by contributing to or staying updated with such initiatives, you can be part of a movement that ensures AI benefits everyone, not just the privileged few. But it's not just about supporting organizations and advocating for regulations. On a personal level, you can educate yourself and others about the implications of AI.

The future of AI isn't predetermined. By raising your voice and demanding change, you can help create a world where AI is a force for good, a tool that empowers everyone, regardless of race, gender, or socioeconomic status. This isn't just a hopeful vision— it's an achievable reality if we all commit to making it happen. By staying informed, supporting ethical AI initiatives, advocating for fair regulations, and educating yourself and others, you can play a crucial role in shaping a future where AI truly benefits all of humanity. Let's embrace this journey with optimism and determination, knowing we can build a more just and equitable world powered by ethical AI. Because now you know how AI and the black box inside AI works.

As we stand on the cusp of the AI revolution, it's crucial to remember that AI is a tool, not a master. It's up to us to guide its development and use to ensure that it reflects our values and aspirations, not our biases and prejudices. By engaging in open and honest conversations about the ethical implications of AI and understanding how it works, we can chart a course toward a future where AI empowers us all as the best version of being a humAIn.

The next chapter will also provide insights into protecting yourself and your loved ones by understanding brain function, especially in the age of AI and misinformation. Understanding these mechanisms will empower you to navigate the complexities of modern life with greater awareness and resilience, keeping you engaged and interested in the subject matter.

Challenge your thinking

- Big companies may discourage AI literacy, as it could lead to individuals creating and using their own AI algorithms. However, this is not a deterrent but an opportunity for personal growth and empowerment. AI bias is not just a concept; it's a reality that can significantly impact your life, regardless of your role as a business owner, employee, or student. It's not a matter of choice; it's a matter of urgency. It's time to understand and act.

chapter 3

COGNITION IN THE AGE OF AI

It ain't what you don't know that gets you into trouble. It's what you know for sure that just ain't so. - Mark Twain

The gorilla in the room

The Invisible Gorilla experiment uncovers a fascinating aspect of human cognition: *inattentional blindness* or perceptual blindness. The experiment yielded surprising results ahead of the adapted analysis for the digital and short attention span age.[26] Back when it was a classroom study, watching a video of people playing with a basketball. They're not playing a basketball game. Five people in white shirts and five in black shirts toss the basketball around the court. Before the experiment begins, study participants are to count how many times the people in the white shirts or black shirts, not at the same time, pass the basketball.

The simplicity of the experiment seems evident after you review the context. However, if you're a participant, you typically don't notice the person in the gorilla suit wandering onto the court, stand for a few seconds, and finish walking through the game. The players continue to play, passing the ball without a look or nod to the gorilla. Once the experiment ends or the video stops, you're asked if you saw the gorilla.

The updated version has black and white shapes sharing a space, moving around, and interacting. Before viewing the scenario, participants must count how often white or black shapes touch the display sides. During the exercise, a red *cross* moves through the middle of the scene, hovers for a few seconds, and continues until it disappears on the other side. The follow-up experiment had a label for programming code cheat conditions.

Around 30 percent of the participants in both tests noticed the Red Cross or gorilla. All of it comes down to selective attention.

Your *inattentional blindness* has parameters preset for the experiment. The same goes for everything in life. You drive every day on the same road; you have certain expectations. If it's regular traffic, you don't notice the rudimentary vehicles. Only when something unexpected happens, a UFO or tap-dancing Sasquatch, does your brain flip a switch and ask, "Did you see that?" The experiment suggests AI, like your brain, already has preset limitations, expects specific outcomes, and won't notice the malware or polymorphic code because it might be dressed in a black or white t-shirt.

In today's digital age, you're inundated with unprecedented information. The volume of data generated globally is staggering, with estimates suggesting that by 2025, we will produce 463 exabytes of data daily. Four hundred sixty-three exabytes is like downloading 212 million years of high-definition video, storing eleven billion high resolution 4K (3840 X 2160 pixels) movies, or holding 463 trillion gigabytes. Imagine fitting the entire Internet into 190 million smartphones or filling 92 million libraries of Congress. The Library of Congress currently has 21 petabytes of digital collection content, comprising 914 million unique files.[27] The human brain, though remarkable in its capacity, has limitations. It can only process a finite amount of information at any given time, and this cognitive load can lead to decision fatigue and information overload. Studies indicate that the average person consumes approximately 34 gigabytes of information daily, yet your brain cannot handle this deluge efficiently.[28] Your teacup had already overflowed long before you picked up this book. Let's explore the limitations of our cognition in the context of data and the role of AI as our lifeguard.

Your brain's epic quest for knowledge:
A mind-bending energy adventure

Do you ever feel mentally wiped out after long hours of doom-scrolling through social media, tackling schoolwork, or just dealing with life's endless information overload? Your brain, that incredible three-pound organ, is on a never-ending quest for knowledge and takes a lot of juice or energy drinks.

Though small in size, your brain is a powerhouse, consuming a staggering 20 percent of your body's resources. Your brain's battery, or about 80 billion batteries, or one for each neuron, could generate enough energy to run a 20-watt lightbulb.[29] Just like a high-performance sports car, your brain's fuel needs increase when it's engaged in complex tasks, showcasing its incredible capabilities.

Envision your brain as a vibrant city, with each thought, decision, and search for information contributing to its bustling energy. The more intricate the task, the longer the journey, and the more fuel you burn. So, when you're immersed in textbooks or grappling with a challenging problem, your brain's prefrontal cortex – the decision-making CEO – goes into overdrive, consuming energy like a parched traveler in the desert. This is your brain's way of demonstrating its dedication to helping you make the best decisions.

Your brain is constantly processing information. But what happens when there's an overload? A massive traffic jam with bumper-to-bumper neurons and you're still trying to get around all that traffic? You're out of luck. Construction crews have the highway blocked; five lanes bottlenecked to one lane of traffic backing up for miles. You'll notice a decline in your ability to concentrate, make decisions, or even remember simple things. This is your brain's way of signaling that it needs a break from the information overload, urging you to manage your mental energy more effectively. Your brain is like a satellite for hunting for information.

Sifting through the intriguing world of Information Foraging Theory, a concept that sheds light on how your brain navigates the complex information jungle, is like trying to find a gorilla in a basketball game while skydiving into Madison Square Garden.[30] It's *exactly* the thing I've done *here* for you to read this book. I've suffered long, painful hours, using a condensed flow of information, investing time and energy to bring you a well-packed version of the necessary information you will need to survive the digital jungle. I did this so you don't have to and for the pay, of course. There are compelling comparisons between your search for information and other animals' quest for food. Wild animals will go to places, even nest near locations where they can easily find food. Your brain is constantly weighing the cost which is time and effort and benefit of various search strategies. Information foraging theory might be an office worker or academic researcher facing recurrent problems with task-relevant information. It boils down to you absorbing information through a streamlined process, so you don't have to go too far to get the data you need for entertainment, learning experiences, talking points at a social event, or a class assignment. You'll use the same gathering tools I used to gather specific information about certain subjects.

Search engines: your brain's GPS

Search engines are like your brain's GPS for information, but not as trustworthy as your noggin. These powerful tools index the vast expanse of the internet, using clever algorithms to guide you to the most relevant information. You already know how to navigate the internet using keywords or characters specified to find particular sites. This saves you time and effort and lightens the load on your energy-hungry brain.

Here's the thing: everything you've searched—images, songs, actor profiles, which one had plastic surgery, which team won—is still there. Search engines are user-friendly databanks utilizing optimized parameters based on what you've input and the outcome. Often, you'll get exactly what you want—review on a local pizza place, apartment listings, ticket prices—but at a cost. Unlike your brain, search engines have a duplicity that many individuals often overlook. Every website you use, even an accidental click, leaves digital breadcrumbs—let's come up with something snappy for the term. I know; let's start calling those tiny files that track your activity *cookies*. I think it could catch on. Every link you've accessed will log your browser history, preferences, and login credentials and store that nifty information as cookies.

Not all cookies are poisonous. Most are yummy! They can help websites remember your preferences, language, and logins, so you don't have to keep telling the site you're not AI because you know what a traffic light is and can click all the boxes, and the AI can't do that—yet. Search engines—all the big ones you can name without me plugging them—have a host of cookies fresh from the digital oven to drop into your computer while hunting for the cheapest chainsaw or favorite hand lotion. Those cookies begin to send and receive data like the aroma of grandma's kitchen, telling the world wide web what kind of lotion you like or your preference for chainsaw manufacturers. Search engines use cookies that help you and they help manufacturers and services to cater to your needs by giving you information narrowed down to specifics.

Search engines have people, and AI is mostly monitoring everyone—everywhere. Your brain has no CFOs, CEOs, or corporate branding. So you can make decisions independently based on a collection of cookies inside your brain. You weigh options and hopefully make informed decisions based on your preferences. Yet, search engine duplicity becomes abundantly clear once you've hunted for that bargain. You will immediately receive product and service advertising based solely on your internet searches. When you look up why Black Ivory is the most expensive coffee in the world and are surprised by the results, you'll immediately get inundated with coffee commercials so they can sell you their product instead of the other brand. Your brain has cookies similar to websites. But you make informed decisions using a collection of data-driven ideas to navigate throughout your day.

Memory is the mother of all wisdom

When it comes to memory, the jellyfish, with its memory span averaging a mere three seconds, is a fascinating creature. Now, consider yourself a unique being with a lifetime of memories shaping your thoughts and actions. The contrast is striking, isn't it? It's a testament to the incredible power and complexity of the human brain, a marvel that we are just beginning to understand.

Your brain is like a high-tech information processing center, juggling two distinct memory systems: working memory and long-term memory. Think of working memory as your mental sticky notes, a practical tool that holds onto information briefly while you use it for immediate tasks. Do you need to remember a phone number long enough to dial it? That's working memory in action, a testament to your brain's efficiency.

But your brain's true treasure trove lies in its long-term memory. This is where your life stories reside—your childhood memories, the lyrics to your favorite song, the taste of your grandma's or websites' cookies. It's a vast library of experiences and knowledge and a repository of emotions that define who you are. Now, expand your memories to active memories, where you can remember a specific detail of an entire day when you were eight years old. Some extraordinary individuals possess the ability to recall an astonishing amount of detail from their entire lives using prodigious memory skills and splinter skills. These memory exercises trigger various neurons storing information that's otherwise unused by most people. Yet even these individuals with Hyperthymesia—highly superior autobiographical memory (HSAM)—have limitations. Your brain is a three-pound universe that processes 70,000 thoughts each day using 100 billion neurons that connect at more than 500 trillion points through synapses that travel 300 miles/hour.[29]

While most of us don't have this superhuman ability, it reveals our brains' immense potential. Our brains are constantly bombarded with information, and storing everything would be like cramming an entire universe into one of those neurons rattling around in your brain.

Information overload: The brain's spring cleaning

So, what does your brain do with all that excess information? It constantly filters through data, deciding what's worth keeping and what can be tossed out. For instance, it might remember your favorite childhood toy but forget the exact date you last saw it. It's like a diligent librarian weeding out outdated books to make room for new arrivals.

Think about it: how often do you remember every movie you watched or article you read? You only retain the information that's most relevant or meaningful to *you*. This forgetting process is crucial for maintaining a healthy and efficient brain.

Scientists have discovered *synaptic pruning* tends to begin when a person is around two years old.[31] By the time they are approximately ten years old, a person's brain will have removed almost 50 percent of the synapses present at two years old. Although neurons comprise about 10 percent of the brain, the rest is cherry gelatin. Actually, "the rest consists of glial cells and astrocytes that support and nourish neurons."

Forgetting isn't just about deleting information; it's a fascinating process of perceiving and interpreting the world around us. In fact, some neuroscientists believe that forgetting is an active process involving specific mechanisms in the brain working to erase certain memories. It's like cleaning the cookies on your computer. For example, proteins such as AMPA receptors weaken synapses, leading to forgetting specific memories. This active forgetting is a marvel of the brain, preventing it from becoming overloaded with unnecessary information and maintaining its ability to function efficiently.

Additionally, your brain uses a process called "reconsolidation" to update existing memories. When you recall a memory, it becomes temporarily malleable, allowing new information to be integrated. This process not only helps adapt old memories to new information but also aids in removing or modifying aspects of memories that are no longer relevant. For instance, if you learn a new fact that contradicts an old memory, the reconsolidation process allows your brain to update that memory with the new information.

Forgetting also plays a pivotal role in emotional regulation.[32] Holding onto every negative experience or piece of distressing information can be mentally exhausting and detrimental to your well-being. But here's the empowering part: by selectively forgetting specific emotional details, your brain helps you move on from past traumas and focus on the present, giving you the control to shape your emotional landscape.

So, the next time you forget where you left your keys or can't recall a minor detail from a conversation, remember that your brain is doing its job. It's busy pruning, consolidating, and optimizing to ensure you function at your best. Embrace the art of forgetting—your brain's natural ability to prioritize and manage information. It keeps you sane, sharp, and ready to take on new challenges, like where you misplaced your smartphone or keys. Our brain needs these forgetting techniques to make unique decision-making every day. Forgetting helps your brain to recharge the next morning to start new information foraging again. Otherwise, you will be too tired to even start another day, leading to cognitive fatigue.

The tired mind sees no options

Cognitive fatigue is the mental exhaustion you feel after prolonged mental activity.[33] This could be anything from studying for hours, working on complex tasks, making daily decisions, and doomscrolling. It's similar to how your muscles feel tired after a long workout, but it happens in your brain.

When planning a vacation, you start searching online for a hotel. At first, you're comparing amenities, locations, and prices across various options. However, after sifting through numerous choices, your brain starts to tire. Eventually, you feel so exhausted that you pick the most prominent option, even if it may not be the best deal. Cognitive fatigue at work is the sheer volume of information and decision-making that wears you down, leading to less optimal choices. It is interesting that while your search may end and you may not purchase the product sometimes, AI will learn from your history and will show you relevant information until you buy. In the next section, we discuss this!

The impact of cognitive fatigue is particularly significant in high-stakes environments such as healthcare. Healthcare professionals often work long hours in these settings and make numerous critical decisions. When cognitive fatigue sets in, it can lead to diagnostic errors, which can be the difference between life and death. Studies have shown that tired doctors are more likely to make mistakes. A tired doctor might misinterpret a patient's symptoms or overlook critical test results, leading to incorrect diagnoses, amputations, misplaced organs, and poor treatment plans. This underscores the crucial role of understanding and managing cognitive fatigue in healthcare and other high-stakes industries.

AI helps combat cognitive fatigue by streamlining the way you access information online. When you browse the internet, small data files called "cookies" are stored in your browser. These cookies record your preferences, search history, and interactions across various websites. AI algorithms use this data to tailor content specifically to your needs, pulling relevant information from other sites. By understanding your browsing patterns and preferences, AI can predict and present the information you are most likely interested in, making your online experience more efficient and reducing the cognitive load required to sift through vast amounts of data. This is the reason you feel someone is listening through your phone or computer.

AI must be listening!

Have you ever thought about a product, perhaps a pair of shoes or a new gadget, and suddenly, as if magic, it appears on your social media feed? It's almost like your phone can read your mind, making mistakes with autocorrect. Welcome to the fascinating world of Information Scent and Information Hook, where AI algorithms play the role of digital detectives, piecing together clues to serve you the most relevant information.[34] We've discussed this a little already with the warm homemade cookies on your computer. Let's dive into how this works and how you can become more consciously aware of these seemingly magical moments so you are not a product of a corporation's AI.

Information Scent is building a skyscraper on the foundation of Information Gathering Theory. When you search for something on Google, you follow a trail of digital scents—keywords, links, and snippets—that guide you to the information you need. These trails help you decide which link to click on based on how strong the "scent" of relevant information is. In the digital world, information scents are meticulously crafted using AI algorithms to capture your attention based on your interests, behaviors, and search history. But with awareness, you can recognize these hooks and make more conscious decisions in the digital realm.

How does it work?

Suppose you're casually browsing the internet, looking at new laptops. You visit a couple of websites, read a few reviews, and add a laptop to your cart but don't complete the purchase. The AI algorithms in the background are taking note of this behavior and getting angry. They understand you're interested in laptops and might be ready to buy soon. This is the Information Scent in action. It's important to note that these algorithms operate within the boundaries of your consent and privacy settings. Next thing you know, you're on social media, and there it is—a perfectly targeted ad for the same laptop you were looking at, maybe even with a discount code to entice you further. This is the Information Scent reeling you in. The AI algorithms have set the bait, hoping to draw you back to complete your purchase.

When you shop online, the paths you take, like browsing

for a new coffee maker, are Information Scents. Later, you see targeted ads for coffee makers on Facebook, X, and Instagram. These targeted ads are designed to entice you to similar products based on your prior activity.

Streaming services also effectively utilize this concept. If you watch a documentary about space on Netflix, YouTube might soon recommend videos related to space exploration, UFOs, and bad streaming service movie recommendations. Here, your viewing history serves as the Information Scent and the subsequent video recommendations act to keep you engaged with related content.

Travel planning provides another clear example. When searching for flights to Paris on Google, you might later encounter ads for hotels in Paris on your favorite news website. Often, search optimization will allow AI to work for you, looking for better flight deals while you're busy streaming bad movies on streaming services. If you set up parameters for flights, you'll get more results than you'll ever need.

Unseen by most, AI algorithms are constantly amassing data from your online activities. These algorithms meticulously scrutinize your search history, browsing patterns, and social media interactions, constructing a comprehensive profile of your interests and preferences. They harness this profile to anticipate what information will pique your interest and when you're most likely to engage with it, a fascinating insight into the world of AI and its ability to decipher human behavior.

However, it's important to remember that these algorithms are designed to keep you engaged and potentially make a purchase, so they might only sometimes show you the most objective information. It's like having a personal assistant who knows exactly what and when you want it, but with a hidden agenda. Don't forget about purchasing that laptop in your online store shopping cart. The AI will keep reminding you about it.

When you visit websites, you often see a pop-up asking if you accept cookies. These cookies are small files that store information about your visit to help improve your browsing experience the next time you return. But if you're concerned about privacy, think twice before hitting "Accept."

For example, cookies can track your online behavior, like the items you add to your cart or the articles you read. This data can be shared with third parties, resulting in targeted ads that seem to know too much about you. A real-life scenario: you search for running shoes, and suddenly, every site you visit shows you shoe ads. You don't have to accept cookies if you don't want to. Click "Reject" or "Manage Preferences" instead. It won't affect your ability to use the site. You might lose some conveniences, like saved login details, but it's a small trade-off for protecting your privacy. Remember, your online experience is yours to control, so choose wisely and browse with peace of mind.

Why you need to be AWARE

Now, it's time for you to take the reins. A fish will never recognize the hook, just the bait. You have the power to understand and navigate these digital hooks and avoid getting snagged. The key is to stay mindful of how your data is being used and to question why certain information appears in your feeds. This understanding empowers you to make informed choices about your online engagement.

Ask yourself:

- Why am I seeing this ad right now?

- What recent searches or clicks might have triggered this content?

- Is this information genuinely helpful, or am I being subtly manipulated?

- Are those cookies in my computer poisonous?

By understanding that these hooks are designed to capture your attention based on your behavior, you can make more informed choices about what you engage with online. Recognize that these hooks are set by companies aiming to bait you into specific actions. Being aware of this can help you navigate the digital landscape more wisely.

So, the next time you see that ad for the shoes you were *talking* about, you'll understand what's happening behind the scenes. AI is always listening. Stay curious, stay informed, and most importantly, stay aware of digital surroundings. You have the power to navigate the digital world with a clear understanding of the AI-driven strategies at play, turning the tables and using the system to your advantage. Once the information hook takes you to the desired site, cognitive anchoring comes into play. Someone starts baking cookies on your computer, and you're eating all of them. This is a psychological phenomenon where your initial exposure to information influences your subsequent decisions. Understanding cognitive anchoring can guide you toward more efficient decisions in the age of AI, whether it's about buying a house, changing a career, or making simple travel plans.

Cognitive anchoring: The stuck mind

Picture yourself in a grocery store, eyeing a box of cereal. The first one you spot is a premium brand for $10. Despite it being way over your usual budget, this initial price, or "anchor," lodges in your mind. Later, when you stumble upon a more affordable brand, say one priced at $5, it suddenly appears a steal—even if it's still relatively pricey. This is the captivating influence of cognitive anchoring. It's a psychological phenomenon where the first information we receive heavily shapes our subsequent judgments and decisions. This 'anchor' sets a mental benchmark that distorts our perception of everything that follows. Encountering a high-priced cereal first can make other cereals seem cheaper in comparison, even if they might not be objectively good deals.

This concept, explained by psychologists Amos Tversky and Daniel Kahneman, reveals that our brains often take shortcuts when processing information. Many decisions are based on beliefs concerning the likelihood of uncertain events such as the outcome of an election, the guilt of a defendant, or the future value of the dollar.[35] Furthermore, experienced researchers are also prone to the same biases when they think intuitively. The fundamentals include coding for AI protocols.

Marketers and advertisers are well aware of this cognitive bias.[36] Marketing effectiveness frequently uses well-seasoned advantages, knowing that presenting a high-priced item first can make subsequent items appear more reasonably priced. This strategy is evident in everything from real estate to retail. Marketing teams have access to real-time data, allowing them or their AI to define how effectively their marketing strategies work. The data must provide prices clearly and concisely for end-users.

The anchoring effect isn't just a theoretical concept; it has tangible implications backed by scientific research. In real estate, listing prices significantly mold buyers' perceptions of a property's value, as documented in the *Journal of Economic Psychology*.[37] Sometimes, the simplest mistake can change the view of the buyer and seller of properties. If a buyer looking for a home between $250,000 and $200,000 won't see listings for houses below $200,000. When you've narrowed search results, you've excluded certain aspects that might otherwise appeal to you or your price range.[38,39]

Our reliance on anchors can lead to several issues. One major problem is biased decision-making. When we anchor on an initial piece of information, we might overlook better options or fail to assess the situation accurately. Overconfidence is another consequence. Anchors can make us feel more confident about our decisions, even when our judgment is skewed.

So, how can you steer clear of the anchoring trap? The key is awareness. When making significant decisions, strive to gather information from multiple sources before forming an opinion. Be mindful of the initial anchors and actively challenge them. For example, if you're considering a major purchase, look at various options in different price ranges *before* setting any expectations or narrowing results. But perhaps the most powerful tool in your arsenal is critical thinking. Question the validity and relevance of the initial anchor. Is that high-priced item really the best benchmark? By engaging in this mental exercise, you can better navigate the information and make more balanced decisions. We discuss a detailed plan on how to employ mindfulness in Chapter 9 in detail, so hang tight!

Cognitive anchoring is a potent force shaping your judgment and decisions, often without you realizing it. Whether buying a car, investing in stocks, or simply making daily decisions, being aware of the anchoring effect can help you see beyond the initial information and consider all your options more clearly. When you focus too much on an information hook, you've already detected the information scent, and the AI knows what kind of cookies your computer likes.

Cognitive anchoring helps to formulate mental models of the universe in our heads. We all have mental models—frameworks that help us understand and make sense of new information. These models act like filters, influencing what we notice, remember, and act upon.

We see things not as they are, but we are

Mental models are the frameworks or cognitive maps you use to interpret and understand the world. These models, such as "the world is a dangerous place" or "hard work leads to success," help you make sense of the vast amount of information you encounter daily by filtering and organizing it based on your past experiences, knowledge, and beliefs. They are the shortcuts your brain uses to navigate complexity efficiently.

Mental models are like the software that help a computer run smoothly, processing data quickly and efficiently. However, just like outdated software can cause a computer to malfunction or crash, outdated mental models can hinder your ability to process new information accurately. A revealing study highlighted the impact of mental models on our perception and actions. It found that people act on only about 30 percent of the information that reaches their eyes. The remaining 70 percent is filtered out, often unconsciously, by these mental models.[40] This selective perception means that much of the information we encounter is not registered or acted upon if it doesn't fit within our existing cognitive frameworks.

To understand this better, consider how we navigate a familiar route, like driving to work. Your mental model of the route allows you to drive almost on autopilot, effortlessly ignoring irrelevant details. However, if significant changes occur, like a new roadblock, Sasquatch, or a UFO, you might miss it if you're not updating your mental model. This concept is crucial in many areas, such as education, business, and personal growth. For instance, in education, students often rely on preconceived notions or outdated knowledge frameworks when learning *new* subjects. New information might not be effectively integrated if these frameworks are not updated, leading to misunderstandings or incomplete knowledge. Often, students between the ages of fourteen and twenty-three think they already know *everything* and why bother learning more? Similarly, companies that need to update their mental models in response to market changes or new technologies can fall behind their competitors. We discuss the urgency of utilizing AI in teamwork in the next chapter for business and teamwork. The consequences of not updating your mental models can be significant, leading to missed opportunities, misunderstandings, and even personal or professional stagnation.

Your brain simulates past experiences to build these mental models, helping you navigate the world. This ability to simulate and predict based on past experiences allows you to react more effectively to new situations. Your mental models are not just abstract concepts but are deeply rooted in your personal history and experiences. They are shaped by the lessons you've learned, the beliefs you hold, and the values you cherish. Understanding this can help you gain a deeper insight into your own mental models and how they influence your perception and decision-making.

So next time you find yourself sticking to old-fashioned thinking, remember: it's like running outdated software. Question your assumptions, seek new information, and update your mental models. This process of updating is not just a necessity but a powerful tool that puts you in control of your understanding of the world. It might be a bit uncomfortable at first, but the rewards of a more accurate and effective understanding of the world are well worth it. Keep your brain's software up-to-date, and you'll be ready to navigate the complexities of modern life with ease and confidence.

The future belongs to those who learn more skills and combine them in creative ways

Ever felt overwhelmed by the sheer volume of information bombarding you every day? You're not alone. Enter AI, your new best friend, for cutting through the noise and making sense of it all. Picture this: when you're creating your own AI, like a personalized GPT, you can program it to focus on the details you often miss. It's like having an assistant who knows exactly what you need, even before you do.

In my research, I discovered that the key difference between novice and expert clinicians lies in their ability to pinpoint exactly where to look. Experts have a knack for ignoring the fluff and zeroing in on the critical info, thanks to years of experience. Conversely, novices tend to get bogged down by textbook knowledge, missing the forest for the trees. AI can bridge this gap by guiding less experienced individuals to the right information faster.

But here's where it gets even better. In the past, designing effective information search strategies was like trying to find a needle in a haystack. We often end up stuck in algorithms designed by social media giants, meant to keep us hooked with endless dopamine hits but not necessarily to help us find what we need. Have you ever noticed how Google or Bing sometimes prioritize sponsored links over the most relevant content? It's like a never-ending game of "Where's Waldo," but Waldo is buried under a pile of ads.

Think about having an AI tailored specifically to you that bypasses all the sponsored noise and gets straight to the good stuff. Your personalized AI could help you search beyond the typical results, providing you with the exact information hook you need. It's like having a search engine that knows you better than you know yourself, helping you easily navigate the digital ocean.[41]

So, next time you're drowning in data, remember AI is here to help. It's not just about reducing cognitive fatigue; it's about enhancing your ability to find and use information effectively. And who knows? Maybe with the right AI, you could start browsing like a seasoned pro or finding that perfect bit of info faster than you can say "search engine."

We comprehend our existence through cognition and consciousness. Understanding how our brain functions can guide us toward leading meaningful lives. In this era, AI integration already empowers you to live a more fulfilling life as a productive individual. With AI as your trusted companion, opening up new horizons for personal growth can happen easily to help you become the next humAIn.

The next chapter underscores the collaborative aspect of working with AI in business, teams, and groups. The economy and the country are built on robust business models that AI supports. You will learn to leverage AI in your business and teams to improve work efficiency.

Challenge your thinking

- If you are a social media company, will you give customers the choice to tweak algorithms?

- Are you open to learning more about AI algorithms and applications that are available to support your cognition?

part II

ADVANCEMENTS

chapter 4

TRANSFORMING
BUSINESS WITH AI

Individually, we are one drop. Together, we are an ocean.
– Ryunosuke Satoro

The Automation

Rick had spent the last thirty years working on the assembly line, taking pride in his work crafting cars. Then, one day, he and the rest of the factory workers received news. The company intended to make changes, replacing many hands-on jobs and giving over to full automation. The transition to autonomous and electric vehicles was the future. The company planned to replace Rick and the rest of the factory workers with AI.

Rick loved building cars. When he graduated high school, his life was about the assembly line, the wrench, and the welding.

His hands understood metal; he felt at home around the scent of oil and understood the perfectly fitted machined parts. Now, the corporation intended to replace him on the floor. They told their workers not to panic or worry too much about stepping aside to let the robots and AI take over the factory floor. The transition would take time, they assured. They intended to keep many workers but gave everyone the option for a buyout or early retirement, an attempt to make the transition seem less daunting.

At 52 years of age, the idea of change terrified him. The company began hosting paid meetings for dedicated employees. The speakers were young and energetic. They knew *everything* about automation and AI but *nothing* about machine operations or assembly work. The factory floor seemed like a foreign land with its whirring and humming automated robots. The sense of loss experienced by the workers as their jobs became irrelevant was profound. Soon, many of Rick's work companions disappeared. Rick's story is not unique.

The auto industry quickly evolved with AI helming the manufacturing floor, transiting traditional assembly line workers out of jobs. As of this writing, Tesla has six Gigafactories across the planet. They are marvels of automation, where AI systems oversee everything from manufacturing processes to quality control. Traditional roles are replaced by tech-savvy positions, leaving many like Rick to find new paths to fight for survival in the robot apocalypse. This shift also brings new opportunities. AI can enhance efficiency, improve safety, and create new job roles. It's not about replacing humans but working together to create a better future.

With Rick's team transitioning into this new era and proper support, they can learn to work alongside AI. Rick has the opportunity to enhance his skills and find new ways to contribute to the factory. People inhabit the workplace, even in futuristic manufacturing plants like Tesla's Gigafactories. They provide a necessary presence to the larger purpose. AI can't do it alone. Automation still needs people like Rick.

This chapter will help you navigate the robot takeover, understand AI for businesses and teams, and unlock the potential to adapt and thrive in this changing world. It's about overcoming fear and moving forward into a future where AI and humanity work hand in hand for a better tomorrow. We will discuss the impact AI has on business ventures. There are future opportunities to create a startup as a single person with a team of AIs and practical action plans to utilize AI to build excellent teams for businesses and projects.

Your new coworker

AI is already transforming businesses, creating increased efficiency, innovation and collaboration opportunities. Humans are
highly collaborative, and we work best as a team and unit. AI
doesn't need bathroom breaks, but it still needs human interaction. AI isn't a replacement for human intuition and creativity.

Marketing teams sift through mountains of data to understand customer behavior. It's tedious and time-consuming, right?
Some firms already employ AI models to reach broader customer
bases and multilayered marketing strategies. Currently, AI generates advertising in real-time based on customer demographics,
improving the ad campaigns in multiple languages with tweaks
that transition user-friendly dialogue.

How about trying to reach a real person in customer service? You're already frustrated with the idea of having to speak
to AI instead of humans. Companies have flipped the switch on

customer service because it's cost-effective to use AI chatbots instead of people. AI communications are reliable and don't try to apologize to you a hundred times during the call because of your issues with the product or service. The likelihood of you speaking to a customer service representative in most companies now is less likely than seeing Sasquatch hitchhiking on your way to work. While it might make you say things to the AI chatbot on the phone you might not actively say to a real person, you will realize the audio prompts you're giving the AI system will eventually resolve the issues.

Many call centers now use natural language understanding (NLU) software. In response to record-high ticket volumes across nearly every industry, many companies accelerated their adoption and usage of AI-powered chatbots to help customers quickly get answers to common questions. Looking back at the medical industry, healthcare providers are using bots to help people get fast answers to questions ranging from vaccine eligibility, availability, tracking and administration, and scheduling appointments.

In other business models, AI doesn't just identify patterns and trends that human analysts might miss; it empowers you to make more informed decisions. AI-driven predictive analytics can forecast market trends in finance, helping businesses make better investment decisions. This predictive power is crucial in today's fast-paced business environment, where staying ahead of the competition often means making quick, data-driven decisions. With AI, we can confidently navigate the complex business landscape.

Ethical considerations are crucial when dealing with AI integration. We've touched on bias and fairness, but it's an ongoing battle with some political agendas that disregard socioeconomic status, gender, and race. Transparency comes back because you must see the system's behavior to understand *why* AI chose certain solutions. Privacy is a big issue for everyone when interacting with AI because the system doesn't know or care about you individually. Misused or mishandled data isn't AI's fault; someone intends to use the system for personal or political gain. When using AI, it's difficult to touch safety factors because programs don't have malicious intent. Of course, if the systems aren't programmed properly due to human errors, AI can't function safely. All of it comes down to explainability. You need to understand how and why AI makes decisions. If the AI algorithms are mysteries, the system needs deciphering to interpret results to understand cause and effect. That means a real need for human oversight with AI. To take the lead in AI technologies, you have to be trustworthy. AI can't be held accountable for its actions. Human error and operation protocols create the landscape to make AI the best it can become.

For businesses AI is already a past
Neighborhood Walmart

Walmart isn't just about rollbacks and everyday low prices anymore—they're diving headlong into the AI revolution, and it's paying off for shareholders. Forget those days of empty shelves and frustrating out-of-stock messages. Walmart's AI is like a super-smart inventory wizard, predicting which products will fly off the shelves and ensuring they're always in stock. It's like having a personal shopper who knows your every whim and desire but with the added reliability of a machine. Walmart's AI isn't just about keeping shelves stocked; it's about getting your groceries to you faster and more efficiently. Their AI-powered route optimization system is like a GPS on steroids, ensuring that every truck and delivery vehicle takes the most efficient route possible, providing you with a secure and reliable delivery service.

However, the AI revolution at Walmart doesn't stop on the shelves. It's also transforming its warehouses, making everything from stocking shelves to picking orders a breeze. It's like having a team of super-efficient robot workers who never get tired or make mistakes. So, next time you're at Walmart, take a moment to appreciate the AI magic happening behind the scenes. It's the future of retail, and it's happening right now at your local Walmart. You'll notice AI at work because you'll have to do all your shopping, including your checkout, because Walmart can't be bothered by hiring cashiers anymore. They want you to do all the work so they can pass on the savings to the shareholders.

Netflix and its algorithm

Despite most personal preferences, Netflix continues to dominate the home streaming services, topping ahead of Max, DisneyPlus, and the rest that are too numerous to list. Netflix overshadows competitors because of its continued aggressive consumer-capturing AI.[42] The AI processes recommendation lists by monitoring user activities and viewing patterns in real-time to determine peak view schedules. The AI-driven data personalizes search recommendations based on movies or shows you watched and how long you watched them. Did you ever binge-watch a season when the system asked if you were awake or had walked away from the TV? It's the AI paying attention to see if you've let the show run in the background while using the bathroom or fixing an egg sandwich. However, it's important to note that users have a vast content library that curves based on preferences. Profiles help filter shows and movies, so if you like true crime and romance— or true crime romance, your spouse might prefer reality or court TV, horror, and sci-fi movies. Following actors, adding 'likes' or 'dislikes' to viewing choices will prompt the system to offer narrowed personalized content, giving you control over the AI's recommendations.

Pfizer's drug development using AI

Envision a world where new medications arrive rapidly, clinical trials run with utmost efficiency without animal testing, and treatments are perfectly tailored to individual needs. This is the future that Pfizer, the pharmaceutical giant, strives to create with AI's help. In this context, AI acts as a super-powered detective for Pfizer's drug discovery process. These advanced algorithms are adept at sifting through mountains of data to identify potential drug compounds that could combat specific diseases, thereby significantly speeding up the discovery process and increasing the chances of finding new treatments. Their collaboration between Pfizer and the Research Center for Molecular Medicine of the Austrian Academy of Sciences (CeMM) has resulted in a new AI-driven drug discovery method that could make it faster and easier to identify small molecules with therapeutic potential.[43]

Healthcare team using AI

The health insurance company Anthem aims to harness an AI-powered approach to healthcare.[44] The company leads the way through messy paper trails, replacing any scrap of human inter- ference by replacing bad decisions with AI-driven approaches. Anthem uses technology infrastructure strategic migration to provide predictive analytics. Anthem uses insurance-funded investments to modernize and build self-healing and resilient healthcare data.

So, Walmart, Netflix, and Anthem—all driven by serving ordinary consumers—use AI to successfully navigate their infra- structure plans. Given the immediate access to all answers every- where, a quick search suggests that 96% of people in the United States own or use smartphones and 21 percent use fitness-based or wearable technology. And 1 in 4 homes use smart speaker technology like Amazon Echo, Google Nest Mini, and Apple HomePod—oh, and Sonos Era 100. So, you wear AI, listen and watch AI, and AI always watches and listens to you. However, AI's true effect will be in ExO companies for future businesses.

What is ExO?

Often, when company leaders receive information about how sales or work decline, they ask why it happens and start pointing fingers instead of looking at facets of the organization that need fine-tuning to appease the shareholders. Sometimes, there are radical shifts in rebranding: Twitter is now X; Dunkin Donuts is now Dunkin; Facebook is Meta; Weight Watchers is now WW. Many people might not know that TikTok wasn't always the Chinese meaning for "shaking sound" or *Douyin* but was once called Musical.ly and rebranded by the Chinese parent company ByteDance. This happens because CEOs come and go, and one power-tie book from the late 70s suggested that if you step into a leadership role, you should first make changes or rebrand. It's not just to shake the money trees; companies use trends in technology to make predictions about what works. Replacing human marketing groups with AI-rive data that cuts through the smoke breaks to get to the heart of the matter: *if we change that, will they buy it?*

The newest version of company additives is an exponential organization or ExO.[45] Ray Kurzweil, Director of Engineering at Google, spearheaded this rebranding technique by suggesting, an organization's ability to leverage new technologies can claim production, output or overall impact that is at least ten times larger than a regular organization in the same field.

Retooling corporate, creative directions kickstarted with a little book published in 2014, Salim Ismail's best-selling *Exponential Organizations* launched the global ExO movement.[46] It contains all the same business models companies had *before* the publication, but Ismail managed to coin *Exponential Organization (ExO)* by immediately investing in rapid growth AI technologies to help companies grow ten times faster than their traditional counterparts. Ismail believes this is the beginning of the fourth industrial revolution, with AI leading the accelerating evolution.

All aforementioned companies, including thousands more, use ExO to expedite data for success. AI has replaced the focus groups to streamline the bottom line. The business landscape is no longer linear, nor is the earth still flat. Exponential change happens because companies upgrade their organizations to information-enabled technology. Companies have turned to and rely on Massive Transformative Purpose (MTP).[47,48] The digital leadership dynamic encapsulates an organization's overarching ambition, representing the fundamental reason for its existence.

A new era of doing business as ExOs

Data retrieval and analysis is a slow, exhausting process, fraught with potential for human error. You've got a company mosaic that's not making sense because you couldn't fit all the pieces together to see the bigger picture. Now, AI is the puzzle master. AI assembles the data puzzle much faster and more accurately. AI is liberating businesses from these tedious tasks, allowing managers to focus on more strategic aspects of their roles. The mosaic is as clear as that CEO bonus check.

AI increases accuracy beyond human capability. It's transforming business operations by replacing outdated methods with sleek, efficient systems. With AI, managers can make smarter decisions and get things done without confusion and mistakes. AI can play its own devil's advocate, presenting challenges and counterarguments and exploring scenarios where the business proposition might not succeed. This comprehensive overview helps the team consider all angles, including human biases. After AI presents its findings, team members can discuss, question assumptions, and add their insights, considering cultural context and emotional intelligence. AI handles the preliminary research and offers diverse perspectives, allowing the team to dive into strategic thinking and creative brainstorming.

In the future, AI will not replace human team members but will complement their abilities. It acts as an intelligent assistant, providing diverse perspectives and cutting through the noise, allowing humans to focus on what they do best. The synergy between human creativity and AI's analytical power could lead to more efficient, insightful, and dynamic team meetings, transforming how you collaborate and make decisions. This optimistic future is within reach with the help of AI.

The fundamentally exciting part is that you can create a company with AI team members and become a successful solopreneur in the startup arena. AI is not just for big corporations; it can also be a game-changer for individual startups, providing the analytical power and diverse perspectives often out of reach for small teams.

Solopreneurs and the future of startups

Peter Thiel, a German American entrepreneur, venture capitalist, and business executive, helped to found PayPal, an e-commerce platform. In his book *Zero to One*, Thiel pushes new ideas instead of mining existing products and services.[49] He suggests going from *0* to *1* means creating something unique, while going from *1* to *n* is just doing more of the same thing. Thiel's ideas gave way to looking at your place in the workspace as a solopreneur.[50] According to Merriam-Webster, solopreneur means "One who organizes, manages, and assumes the risks of a business or enterprise without the help of a partner." Much of this solo entrepreneurship is easy to think about, but is it practical in the everyday world of employment? It's easy for Thiel to suggest these grand ideas and implement company changes directly affecting the bottom line because it's *his* bottom line. But you could use the solopreneurship mindset in every aspect of business today. It comes down to taking ownership of your actions in a working environment.

Why not look at making a living using solopreneurship as a model to build success? You can manage every business aspect and not rely heavily on delegating tasks. If you're the creator and founder of your business, you're likely solely responsible for the company's success. Sometimes, solopreneurs need help but don't want to hand over too much of their business to people they can't rely on or don't know directly. They will use contractors or freelancers to perform specific tasks but not take part in daily operations.

Often, entrepreneurs use their venture capital to start business plans and hire outsiders to run daily operations. They expect quarterly earnings and profits and someone to answer for any weakness in the business. As a solopreneur, you don't have to outsource any of your fulfillment obligations if you don't want to hire staffers.[51] However, depending on your business, product, and services, you will wear many hats. That might get your foundation going, but if you're not utilizing the right AI tools, you'll spend more time fixing problems instead of running a successful business.

Unlike entrepreneurship, which throws money at different things to see if it sticks and pays off, solopreneurs seek a specific niche that offers focused services and builds a solid customer base to keep that single entity profitable. It takes research and time, but eventually, you'll find a marketing strategy that allows you to create a working business model ahead of launching your solopreneur. Brand consulting, social media management, and book promotions are creative services based on individuals focusing on single entity endeavors that allow solopreneur sponsorship. Eventually, you'll need to grow your team. But if you're already moving forward with AI services working tirelessly in the background, you'll know how to use the technology to seek niche opportunities to exploit.

The significant difference between solopreneurs and entrepreneurs lies in their business scale goals. Entrepreneurs aim to upscale their business venture, make it glossy and profitable, and eventually sell it to investment firms. In contrast, solopreneurs intend to keep a profitable business independent, continue running it, and make a tidy living off something that makes them and their customers happy.

As a solopreneur, you know that financial responsibility won't be as risky as it is for entrepreneurs. Venture capitalists hire outsiders to run the business. They own the company but have employees who might be as strong-minded about the business as the entrepreneurs. Yet, in the end, they suspect the entrepreneur will sell to the highest bidder for the right amount of money.

Solopreneurs have next to no employees and very little overhead because you're housing the business model inside a single-member LLC or sole proprietorship. Those licensing options allow for seeking grants and are simpler to manage. Your employees live inside the digital world of your high-end computer. If you've got a niche market for physical products, you'll need space for storage and possible working accommodations for a few employees. You might have no trouble working out of a closet or basement packaging and shipping products. However, your employees need certain comforts that require overhead costs.

Successful solopreneurs begin their endeavors through research and marketing strategies, often using AI features designed for business and niche models. Podcasters, finance experts, and authors are traditional solopreneurs who use basic ideas to build dreams. Web designers, event planners, furniture restoration, photography, and graphic design: These growing markets continue to help people like you succeed in solopreneurship. Today's successful business ventures utilize AI to run frontend and backend jobs, allowing solopreneurs the freedom to run their businesses without the overhead capital needed in traditional settings.[52]

Now, let's look into some practical approaches where you can use AI for your business and teamwork. These strategies were derived from recent successful business strategies.

The four-step action plan for small businesses and teams for utilizing AI

The four steps are SAIL.

S- Simplify teams data detective work

A- Automate the boring stuff

I- Integrate AI insights

L- Leverage AI to build a dream team

In today's fast-paced business world, the S.A.I.L. approach—Simplify, Automate, Integrate, and Leverage—offers a robust framework for seamlessly integrating AI into your operations. Whether you're leading a team or embarking on a solopreneurial journey, AI can transform your workflow, making tasks more efficient and insightful. AI coaches teams, enhancing communication and productivity, while solopreneurs benefit from automation and data insights. However, solopreneurs should consider consulting services for specialized support. Adopting the S.A.I.L. strategy allows you to navigate the complexities of AI integration and steer your business toward success. Whether you are a large corporation or a small business, you can utilize these concepts of SAIL to improve operational efficiencies.

Step S: Simplify teams data detective work

Don't let the volume of data intimidate you. Start with a specific, burning question you want to solve. Perhaps you're wondering why certain products are more popular or want to identify when your website traffic peaks. Focusing on a specific problem will make the process more manageable and yield results faster. It's like solving a puzzle one piece at a time rather than trying to piece everything together at once. Every puzzle has key factors that bring a clearer picture.

If numbers and spreadsheets aren't your thing, don't worry. Some experts live and breathe data, others partner with a data analyst or consultant who can help you decipher recommendations for the perfect AI tools you need. You can use as much as little as you need for research. Much of the hard work already has a platform to glean your solutions: YouTube. Influencers aren't always your trusty guides, but you can take snippets of valuable tips, turning complex problems into actionable insights. Using free services ahead of paying for serviceable actions typically means spending a little more time researching, but you'll get more from the knowledge that you might get paying for services.

Embrace the mindset of experimentation. AI is all about learning and improving. Think of it like a science experiment – you try different things, see what works, and adjust your approach. Don't be afraid to test various AI tools and techniques, such as machine learning algorithms, natural language processing, or predictive analytics, to discover what provides the most valuable insights. Each experiment is a step closer to mastering and leveraging your data to make smarter decisions. Your brain will expand with every nugget of information pouring into your eyes and ears. AI can sort and process the hard parts so you can get the most out of the research.

Remember, AI isn't just for tech giants like Google or Amazon. Many sellers utilizing Amazon or Etsy either come from or move to private websites to lower their overhead costs and pass on the savings to customers. They use AI to make better business choices and give you valuable insights. You make better decisions and ultimately achieve more tremendous success by embracing the robots living inside the machines. Of course, there are potential risks, such as data privacy concerns or the need for continuous learning and adaptation. Along the way, you'll find those critical pieces making up the secure borders, firewalls, and VPN gateways to bring success.

Step A: Automate the boring stuff

Building your solopreneurship can be daunting. You might feel tremendous pressure to increase your business success immediately. However, you can unlearn what's expected and build a better business model with AI. Don't forsake the foundations of traditional management strategies. AI can improve profitability, reduce costs, and reduce unplanned downtime. And remember, AI doesn't take coffee breaks, providing you with reassurance and confidence in your decision to incorporate AI into your business.

With AI, you can easily manage physical and virtual machines, saving you valuable time. The beauty of AI lies in its ability to access and process vast amounts of data, allowing you to streamline your automation goals. You can use specific programs to automatically generate spreadsheets, crawl websites, and extract information from online sources. AI can swiftly debug and even identify faulty code in your source tools.

Generative AI can handle the essential background operations, freeing up your time to explore open-source materials that align with your business needs. By implementing effective automation, you can continuously enhance your business model, with real-time data constantly improving your solopreneurship.

Step I: Integrate AI insights

AI already helps businesses shape their decisions. Pieter Den Hamer, Senior Director and Analyst at Gartner says, "With continually more dynamics and complexity in modern-day business—especially digital business—our capabilities must improve to make the best possible decision in the shortest possible time, in a scalable, risk-conscious, consistent, adaptive and personalized fashion." He then takes a big breath to continue, Moreover, the decisions that we make today can't be biased or based on yesterday's situation awareness; they must reflect the here and now.[53]

It is nearly impossible to do business in a digital domain without AI automation, allowing for fast, accurate decisions happening faster than you can blink. AI has the ability to analyze large datasets without error. Your decisions based on AI learning only complement human decision-making. The levels of AI influence can assist in basic automated tasks. Augmented AI allows for a two-way process aimed at machine learning from human input, using their insight to improve AI decisions. Automated AI allows for a complete process of technology replacing human interference. Once you've implemented the automation in your business, AI can finally take over the world.

Step L: Leverage AI to build a dream team

We all will love a workplace where communication flows effort-lessly, conflicts are extinct, and collaboration is a breeze. That's the kind of dream team AI can help you build. Think of AI as your team's coach, providing insights and tools to strengthen bonds, boost morale, and skyrocket productivity.

Building a high-performance team focused on cultivating a culture of trust and transparency shouldn't take any effort. If you've expanded your solopreneurship to include other people (not robots), encourage open communication, allowing everyone to feel safe sharing ideas, concerns, and mistakes. AI can assist by analyzing communication patterns and identifying potential issues before they escalate. For instance, AI can flag a sudden decrease in team interaction, which might indicate a problem beyond performance. Your team might not see the solution for a change, but AI can provide examples to simplify problems, making solutions easier to convey. Next, embrace the convenience of AI collaboration tools. These platforms, like Slack and Micro-soft Teams, serve as a virtual headquarters. They offer AI-powered features streamlining communication, automating tasks, and even translating languages in real time. It's like having a universal translator and a project manager all rolled into one, ensuring smooth and efficient collaboration.

Finally, celebrate wins and learn from losses by empowering AI to handle mundane tasks that improve business performance. Trello, Monday, and Asana integrate AI to monitor project time-lines, suggest optimizations, and predict potential roadblocks. AI always knows how to motivate your team and keep them on track. By leveraging these AI tools, you can boost your team's performance, foster a positive work environment, and drive continuous improvement. By embracing AI as your team's coach, you'll create a more positive and productive work environment and unlock the full potential of your team's collective intelligence.

Ditch the losers motto: This is how we do it

Recent studies show that businesses that fail often share a common trait: a fixed mindset encapsulated by the phrase, "This is how we do it." This mentality can spell doom in an ever-changing world. Competitions are increasing, and you need to adapt and evolve to thrive. If you catch yourself telling your team, "This is how we do it," it's time to think again. Holding onto outdated methods without room for innovation is a sure path to extinction. Embrace change, and you'll be more likely to keep your business alive and thriving.

Think of AI as your secret weapon. It offers an unprecedented opportunity to learn, grow, and enhance your business operations. Be open to new ideas and willing to take risks. Sticking to the old ways might have worked once, but it's a recipe for disaster in today's fast-paced world. As a leader, fostering an environment where new ideas are welcomed and experimentation is encouraged is crucial. Imagine your team suggesting innovative solutions and feeling empowered to explore them. This kind of dynamic thinking keeps your business relevant and competitive. It's like having a safety net against the ever-looming threat of becoming obsolete. If dinosaurs had access to AI, perhaps they wouldn't have gone extinct! Don't let your business be the dinosaur in a constantly evolving world. Instead of saying, "This is how we do it," say, "Let's find a better way."

Encourage your team to challenge the status quo and explore new technologies. Remember, fear of change is the enemy. Embrace AI and other innovations to stay ahead. By doing so, you're surviving and thriving in the competitive business landscape. So, ditch the old mantra and adopt a mindset of continuous improvement. Your future self—and your business—will thank you for it. By embracing AI and integrating it into your team's workflow, you're not just keeping up with the times but setting the pace for future innovations. This forward-thinking approach will ensure that your team not only adapts to the changes AI brings but thrives on building the humAIn team.

The next chapter explores groundbreaking ways education will evolve, impacting us, our children, and future generations. Get ready to discover how AI will transform learning, making it more personalized, efficient, and accessible.

Challenge your thinking

- Have you ever imagined the potential of an AI-based system for employee evaluations in your organization? Traditional evaluation methods can be prone to human bias, making the process unfair and inconsistent. However, an AI supervisor can provide a more objective, data-driven assessment of employee performance, ensuring that evaluations are based on merit and actual performance metrics rather than subjective opinions. This could be a game-changer for your organization, paving the way for a more objective and fair evaluation process.

- As a solopreneur, the potential of leveraging AI bots could be a game-changer for your business. Do you know which AI tools and bots are currently available to streamline your operations? From customer service chatbots to marketing automation tools, a wide range of AI solutions can help you manage tasks efficiently, allowing you to focus on strategic growth. The possibilities are endless, and the future of your business could be more exciting than ever.

chapter 5

REINVENTING EDUCATION WITH AI

Those who can, do; those who can't, teach.
– George Bernard Shaw

The dream to teach

Emily always knew she wanted to be a teacher. Growing up in a small town in Colorado, she was inspired by her own primary school teachers who made learning fun and exciting. Her dream was to bring that same joy to her own classroom one day. However, she quickly learned that the path to becoming an educator was fraught with financial challenges.

After high school, Emily was accepted into a reputable university's education program. She was thrilled, but the excitement was short-lived once she saw the tuition fees. With her family unable to help financially, she turned to student loans. By the time she graduated, Emily had accumulated $60,000 in student debt.

Her first teaching job in Denver came with a starting salary of $40,000 a year. While it was a dream come true to finally be in her own classroom, the reality of her financial situation hit hard. Emily's monthly loan payments were $600, a significant chunk of her take-home pay. Her rent was $1,200 a month, leaving little room for other necessities. Every month was a struggle to make ends meet. She often had to choose between paying her student loans and buying groceries.

Emily's financial stress was relentless. She picked up part-time jobs during the school year and over the summer, working as a tutor and even as a barista at a local coffee shop. Despite working long hours, she still found herself falling behind on her loan payments. The interest kept accumulating, and her debt seemed to grow instead of shrink.

Her dedication to her students never wavered, even as she juggled multiple jobs. She poured her heart into teaching, creating engaging lesson plans and spending extra time with students who needed additional help. Yet, the constant stress of her financial situation began to take a toll on her health. Sleepless nights became common as she lay awake worrying about her bills.

Broader implications

Emily's situation isn't unique. Countless teachers across the country grapple with similar financial issues and the urgency of the student debt crisis cannot be overstated. Depending on the location, some teachers must take second jobs or leave the profes-

sion altogether, depriving students of dedicated and experienced educators. The broader implications of this issue are deeply concerning, particularly when considering the impact on the quality of education. As of 2024, Americans owe a staggering $1.727 trillion in student loan debt, with the average federal student loan debt at $37,056. This crisis affects millions of individuals, with over 43.2 million federal student loan borrowers struggling to manage their debt and accumulating steep interest rates.[54] The financial burden circling educators created desperation in schools across the country.

This chapter will enlighten you with alarming facts about the decline in educational standards and the economic and logistical reasons behind it. However, it also presents a ray of hope. Many dedicated professionals understand the burden of debt associated with proper educational needs to get what they want out of employment endeavors. We'll look at how merging technologies with life goals can mean getting an education and not starving to pay back loans. When used mindfully, AI can help us navigate challenges and enhance the quality of education, thereby supporting the next generation. It offers a promising path forward, inspiring hope and optimism in the face of these obstacles.

Education system creating inequality

I know this may sound so absurd as you may have thought education was supposed to teach equality and create a just society. However, you may need to look at the history and the system we have created so far. The current education system in the USA fosters inequality and hinders the creation of a balanced society. As mentioned before, in many areas, the quality of education children receive is directly linked to the condition of the local community. Higher property taxes in affluent counties fund better schools, while lower-income areas struggle with underfunded schools. This disparity makes vast differences in educational opportunities and outcomes, perpetuating a cycle of inequality from very young ages. Not to depress you with facts, but it's vital to understand some roads have better pavement than others. The hyperbole about student loans isn't new and discouraging for college-age students trying to get a bearing on their lives. Dedicated teachers are still necessary. Yet, fewer people want the financial burdens linked to higher educational standards. And there is no guarantee the location you end up as a teacher will fill your bank account.

The questions surrounding educational inequalities often concern cost-effective measures set up to fail in many urban communities. Social classification is the most significant predictor of academic success in traditional education. Performance gaps in social status, especially in early years, set examples for children who fall behind, often stay behind and rarely gain traction from the lost ground.

Children's socioeconomic status (SES) has steadily declined since the mid-2000s. Now, that doesn't mean we have feral children running amuck on late-night city streets. But it does mean that the current educational curriculum isn't working for many Americans. Much of the trouble with education standards falls from the top with no one's fault based on their income levels. There is a societal failure in the undeniable relationship between education and economic inequalities.

School funding has been the source of most disparities in American equality. The school experiences of African-American and other minority students in the United States continue to be substantially separate and unequal. This inequality is a stark injustice that needs to be addressed. Few Americans realize that the U.S. educational system is among the most unequal in the industrialized world. Students routinely receive dramatically different learning opportunities based on their social status. In contrast to European and Asian nations that fund schools centrally and equally, the wealthiest 10 percent of school districts in the United States spend nearly 10 times more than the poorest 10 percent, and spending ratios of 3 to 1 are common within states.[55] Poor and minority students are concentrated in the least well-funded schools, most located in central cities or rural areas and funded at levels substantially below those of neighboring suburban districts.

Looking deep into it

Formal education has shaped the foundation of student lives throughout U.S. history and worldwide. Think back to your school days. You remember there are at least one or two teachers between kindergarten and twelfth grade. There is one educator who stands out because of some impact they had on your life. You're thinking of a name right now.

That one teacher who connected with you on a deeper level than any other, and their impact on your life is a testament to the power of education. That one teacher, possibly unknowingly, steered you on the course you took to better yourself. Someone believed in you and gave you directions to the road you're on now. You're reading this and it didn't happen by magic. But what happens when there are no more teachers? In 2024, a new crisis emerged: a severe shortage of educators. This is not just a number on a statistic. It's a threat to the quality of education. The impact on our students is profound, with embattled administrators, over-burdened counselors, and burned-out teachers struggling to meet their needs.

Most senior educators exited schools during a little thing that happened to everyone on the planet a few years ago: the global pandemic. This crisis significantly disrupted the education system. Nowadays, fewer students want to become teachers when they leave high school. Jay Schroder, a veteran teacher and author of *Teach from Your Best Self*, states, Current teachers are now unlikely to encourage young people to enter a profession that is as challenging as teaching has become, And it's not only teachers who are discouraging the young people they care about from becoming teachers. A 2022 nationally representative survey found fewer than 1 in 5 Americans would encourage a young person to become a K-12 teacher.[56]

Every school nowadays has highly challenging conditions. Not every educator can be a hero and adversity is never-ending. Currently, some college students earn education degrees but decide *not* to pursue teaching licenses. Others receive teaching licenses *without* college degrees—a practice contributing to the shortage of *qualified* educators.[56] In some states, educators only need to show up to get paid, a situation that is not conducive to attracting and retaining high-quality teachers. They're using outdated curriculums to prepare students for *testing* instead of educating them for life events. If the warm body sitting at the head of the class looks like a teacher, hands out assignments, and keeps everyone in the classroom from eating glue all day, they're considered educators. Don't worry about the fact they don't have a teaching license. As long as they're not felons and maybe can pass drug tests, that's good enough for some schools. There are potential solutions, and with collective effort, we can overcome this shortage.

My kid is going to the BEST school! Are they?

Let's talk about a significant issue that's affecting everyone. You know how we obsess over getting into the best school districts? We spend a ton of money on education, but guess what? U.S. students still rank behind many of their international peers in math, reading, and science. This underperformance not only affects our students but also significantly impacts our economy. U.S. students rank 25th in science, 30th in reading, and 36th in math.[57] More than 30 countries outperform U.S. students. We can take diversity off the table when discussing educational standards worldwide. In a recent fact-check from the Organization for Economic Cooperation and Develo Economic Cooperation and Development, the millennials in our workforce tied for last on mathematics and problem-solving tests among the millennials in the workforces of all the industrial countries tested. We now have the worst-educated workforce in the industrialized world.[58]

The educational discrepancies between the U.S. and other countries relate to the cohesion between schools and the work-force. Remember the incentives students in Shanghai, China, get when they better themselves through education? The same obviously doesn't apply to U.S. students. Common Core State Standards (CCSS)—a set of high-quality academic standards in mathematics and English language arts/literacy (ELA)—no longer applies to U.S. schools.[59] The standard failed because teachers were being judged against student performance on tests that did not measure what the teachers were supposed to teach, there were no curriculum materials available to support what the students were supposed to learn, the teachers had never been taught to teach what their students were supposed to learn, the way students progressed through the grades had not been redesigned against the targets specified by the standards, and no effort was made to reorganize the work of teachers so that they would have more time for students who would need additional help to reach the standards. You might pay top-dollar taxes for school districts, but are the educators in your school district the best influencers on your kids?

Now, let's turn our attention to research as we are often supposed to teach in higher education, where we innovate, and that advantage has made the USA one of the leaders in the world.

Research infrastructure in the USA compared to the world

Despite challenges, the United States still boasts the best research infrastructure globally. The massive amounts of tax dollars funneled through agencies like the National Institutes of Health (NIH), the National Science Foundation (NSF), and the Department of Defense (DOD) ensure that we remain at the cutting edge of scientific discovery. The NSF reports, "The United States (27% or $656 billion) and China (22% or $526 billion) performed about half of the global R&D." However, Switzerland spent 3.36% of its GDP on research and development in 2021.[60] But there's a growing problem within this system that threatens its very foundation as oftentimes we hear "the system is rigged".

I've had the privilege of being on an NIH review panel for giving grants to other researchers and let me tell you, what I've seen is troubling. During the grant review process, the system isn't blinded. Reviewers know exactly who submitted the grants and from which universities they come. This opens the door to bias. Picture this: a senior reviewer, aware that a grant is from a friend or a prestigious institution, subtly or not so subtly influences the other reviewers. They say glowing things about the grant, perhaps because of personal connections or institutional biases, and just like that, the grant score is manipulated. This manipulation isn't just a minor hiccup—it's a systemic issue. It leads to the funding of less innovative or even mediocre research, while truly ground-breaking work from lesser-known institutions or researchers might get sidelined. Over time, this erodes the integrity of our research output. A study has shown that 80% of research is not reproducible.[61] Another recent news article published in 2023 stated that more than 10,000 research papers were retracted due to fraud.[62] This is alarming because reproducibility is a corner-stone of scientific credibility. When research can't be replicated, trust in science diminishes. People begin to doubt the findings, and the very essence of scientific progress is called into question. "The system is rigged"- we heard that for other fields and unfor-tunately, the poison also got into research. The federal agencies, who get paid from your tax money, are apparently in a state of denial and only serving their own purposes. As misinformation is spreading, how can a society be strong if people lose trust in sci-ence and reason? This is a challenging time due to the constraints and lack of government initiatives and awareness.[63]

The dire need for change

So, let's talk about the compelling option of trading in traditional university education for the AI revolution. You already know going to college is costly. It's not about making your parents happy when you graduate from a University. It's not about paying steep student loans later in life when you decide the dead-end job you have now is worth keeping. Everyone learns differently. Without sounding too cliché, the future of AI is already here, and it's more relevant than ever. It's practically in your bones. But aside from the cost, nowadays, getting that diploma in general degrees doesn't get you anywhere. High-paying jobs are right around the corner, and embracing AI learning could be the proper cost-measured education you need to get ahead in life.

The for-profit education system has quietly and systematically created a niche, making everyone believe if you want to succeed in life, you must get a "proper" education through long hours of classroom studies, suffering through 'core curriculum' classes you will likely *never* use while swimming deep water to reach for that 2-year or 4-year degree. Concurrently, you'll feel the heavy drag of the undercurrent because of the possible need to get over your head financially to pay off those student loans once or if you come out of the deep water holding an associate's or bachelor's degree.

What universities won't tell the public is the fact they know their brick-and-mortar days of education are coming to an end. Most universities nowadays will not keep tenured professors anymore. Colleges and universities now keep a staff of adjunct 'instructors' instead of keeping dead-weight professors who lost their way and burned out long before you got through their syllabus. Essentially, they've replaced every retired tenured professor with long-term substitute teachers they can terminate on a whim instead of paying for aged intellectuals. Higher education centers already use hybrid classes where you learn online and have a day or two in lecture halls.

Online universities began popping up almost as soon as the internet flourished. Still, brick-and-mortar universities told employers, "See, we don't care if those students stay home to save money, take classes online, and pass their finals." What you don't understand as an employer is the fact those 'online universities' aren't accredited schools. This is a government-sanctioned term cooked up by brick-and-mortar education factories that say that the degree you earn online is no good. So, for some reason, employers looked at your degrees earned online as nothing more than coarse toilet paper. All that happened because you can't get student loans for online classes, only brick-and-mortar colleges. Why is that?

Trade schools took off in the late 1970s. Anyone who needed focused training for heavy industrial labor could earn their certification through trade school. While for-profit universities had thumbed their noses at trade schools, they had convinced the Wall Street bozos trade school wasn't for education. Yet, those people who were rich failed to understand that it took those people in trade schools to build those skyscrapers. They used to look down on the rest of the people because their 4-year degree on the wall in the corner office came from a 'traditional' brick-and-mortar university.

AI isn't replacing brick-and-mortar for-profit higher education.[64] It's already here and integrated into those universities. Every new semester brings AI deeper into the learning annex. You are required to have computers or access to computers as soon as those colleges accept you. It means you're paying for an education that relies on AI to handle some of the workload. Many older educators fear being replaced by AI. But don't let them fool you; they're afraid of being replaced by younger educators, too. Recently, the COVID pandemic sprang up, causing a massive disruption in education factories.

To save our students, most schools turned to AI to help fill in the gaps for continued education. Governments scrambled to figure out how students could continue learning while not sitting in classrooms secretly playing games on their smartphones. So, AI had already saved the world.

To improve research dollar spending, federal organizations need to change how they review grants. It should be blinded so reviewers do not get influenced. The grants should be evenly distributed between researchers. Currently, the ones who play the system get multimillion-dollar grants, and many prominent and meritorious scholars are leaving academia to go to industry. Your voice matters as it is your money, and it should not be wasted on playing the system. AI should be used to match grants with reviewers instead of the manual process ending up with bias and nepotism. Research should inform teaching to educate the minds of our new generations.

It is the supreme art of the teacher to awaken joy in creative expression and knowledge

AI cannot and likely will never replace the necessary human connection associated with learning. Personalized learning paths for students dictated by computer-dispensed knowledge started as soon as a teacher decided to use a computer to draft a test sheet and a copy machine to print off the papers. Humans build meaningful relationships with mentors.

The U.S. Department of Education's Office of Educational Technology already stresses the importance of integrated AI in classroom settings. The curriculum requires educators to be present in the classroom with students in AI environments. Administrators receive constant feedback through the integration of allowing everyone to understand the impact of AI on students. Schools require government funding to stay open. All academia needs a return on their investments. Parents can get real-time progress reports from AI integrations. The technology involves monitoring as much as the students need mentors and teachers to guide them through the learning process. AI isn't coming for your jobs; AI will make them more manageable.

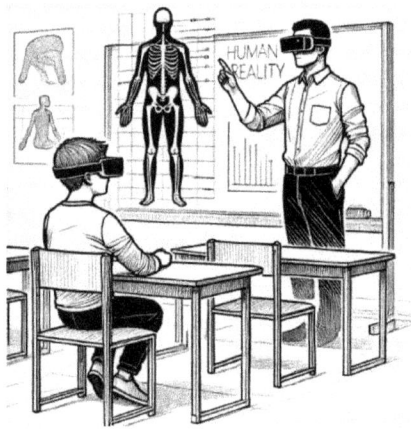

Allow students to explore AI on their own. Whether engaging with chatbots, testing generative AI, or using other AI-powered tools, these experiences help them develop critical AI literacy skills. For instance, a symptom checker app powered by AI can help students understand how AI works in everyday field operations, making the technology more relatable and easier to understand. AI classroom experiences already exist in the pockets and backpacks of children worldwide. If you bring it forward, making it an open classroom experience, the children already embracing the technology will benefit from positive teacher reinforcement.

Augmented Reality (AR) and Virtual Reality (VR) are

revolutionizing classrooms, making learning fun and engaging for students. Imagine students exploring ancient civilizations, stepping into the shoes of historical figures, or diving (safely) into the ocean's depths—all from their classroom. AR and VR applications like Google Expeditions and zSpace are already used to bring such experiences to life, allowing students to explore and interact with 3D models and virtual environments.[65] For example, biology students can dissect virtual frogs or examine the human heart up close, providing an immersive learning experience that traditional textbooks can't match.

Building AI literacy for career choice

AI continues to transform the job landscape, handling some roles and creating shifts similar to those seen throughout human history. Take the Industrial Revolution, for example—carrier drivers had to adapt to new roles as industries evolved. Similarly, AI handles many low-level programming tasks while new job openings emerge, particularly in areas requiring human intuition and creativity.

AI excels in tasks with defined rules and patterns but needs to work on the nuances and unpredictability inherent in many specialized fields. Domain expertise remains irreplaceable. For instance, as an infectious disease physician with thirty years of experience, you understand that evidence-based medicine is limited in complex cases. AI can support protocols but needs more experiential wisdom to navigate every unique scenario. Your experience and nuanced understanding are invaluable, and no AI can replicate this depth of knowledge.

When considering your career path, aim to develop domain expertise. Jobs centered on deep, specialized knowledge will be less susceptible to automation. Look for educational programs integrating AI into their curriculum, regardless of the field. Even in areas like philosophy, programs should teach you how to use AI tools to enhance your understanding and work. For example, a business analyst using AI tools can use vast data to find trends. Still, your insight and experience will interpret these trends meaningfully for strategic decisions. AI can assist but cannot replace the human touch in understanding and applying knowledge contextually. If you're on the path to world domination, you'll need AI in your corner more than henchmen protecting your secret volcanic lair.

Children are not things to be molded but are people to be unfolded

Our children must achieve certain milestones to succeed— graduating from a prestigious university, landing a high-paying job, or following a career path we envision as the best course for their futures. However, history shows us that the individuals who genuinely changed the world were often supported by parents who encouraged their dreams, no matter how unconventional they seemed. For instance, Thomas Edison, whose mother, Nancy Edison, pulled him out of school after his teacher labeled him "addled." She decided to educate him at home, nurturing his curiosity and allowing him to explore his interests. Edison became one of the greatest inventors in history, giving us the phonograph, the electric light bulb, and countless other innovations. Imagine if Nancy had forced Thomas to follow a conventional path instead of supporting his unique learning and thinking. Tesla was a failure in school. Albert Einstein struggled in school and often clashed with his teachers. His parents recognized his potential and supported his independent learning style, providing him with books and resources to explore physics and mathematics on his terms.

Instead of pushing children toward what we perceive as a *guaranteed* path to success, we should focus on helping them find their flow—where they experience effortless joy and deep engagement in activities they love. Psychologist Mihaly Csikszentmihalyi, who coined the term "flow," describes it as a state of complete immersion and satisfaction in what one is doing. This state is not only associated with high productivity but also with genuine happiness.[66]

Encouraging your children to pursue their passions doesn't mean letting them wander. It means guiding them to explore different interests and supporting them in finding what makes them truly happy. It means recognizing that traditional milestones do not solely define success, but the joy and fulfillment one finds in one's pursuits.

Remember that the most remarkable achievements often come from those who dared to dream differently and were supported by those who believed in their potential. Let your children figure it out. You provide them with the tools, encouragement, and freedom to discover their flow and watch as they create their paths to success and fulfillment. Your role is not to chart their course but to support them as they navigate their journey. Be reassured that you are paving the way for their success by supporting their unique learning and thinking styles.

Power of thousand teachers to one student

In the last few centuries, education has undergone a profound transformation. Initially, education was a privilege reserved for the elite, where a single teacher would provide personalized instruction to the sons and daughters of wealthy families. This model evolved dramatically, leading to the current system where one teacher educates 50, 60, or even 100 students at a time. In the United States, the average teacher-to-student ratio is about 16:1, while globally, it varies but often leans towards larger class sizes in many countries with fewer teachers.[67]

You remember the challenges of navigating the education system yourself. Sometimes, teachers and instructors didn't have the time needed for a more one-to-one experience for specific questions. With AI, each student can receive personalized attention akin to having a thousand teachers focused on their needs.

AI monitors how long students engage with problems, identifies where they struggle, and provides immediate, tailored support. This technology can transform education from a one-size-fits-all model to a personalized learning experience that adjusts to each student's pace and style.

For instance, consider homeschooling. Previously, it was challenging due to the need for consistent quality monitoring. AI can track a child's progress in real time, observing how they interact with materials, how quickly they read, and how well they comprehend the content. This continuous, personalized feedback loop helps identify and address weaknesses in a student's study plan, making homeschooling a viable and effective option. Such an AI-driven approach is revolutionary and essential for the next generation. Teachers, constrained by time and resources, cannot always provide the personalized interactions students need. AI offers a solution by providing a customized tutor for each student, ensuring no one falls through the cracks.

Think of the potential here: every child has access to thousands of AI-powered teachers, each dedicated to understanding and improving that child's learning journey.[68] This technology can help students realize their full potential, fostering a love for learning and curiosity that traditional methods often stifle. The future of education is bright with the integration of AI, and we must embrace this change. By leveraging AI, we can create a more equitable, efficient, and effective educational system. This will help students excel academically and prepare them to thrive in an increasingly complex world.

The purpose of education is to replace an empty mind with an open one

We all know students today get a little help from ChatGPT or Gemini. The big question is: Can we stop them? Of course not, nor should it be an issue. Fundamentally, if AI flourished a hundred years ago, you wouldn't be reading this book. You'd absorb the information through osmosis.

Education aims to help people gain skills to contribute to society and advance knowledge for the next generation. Historically, education started with religious teachings. Fast forward a few centuries, and we've turned it into a business. Students used to need classrooms because there was no other way to learn. Now, with all the info on YouTube and other platforms, do we need to cram facts that are a Google search away? Shouldn't we teach them how to apply that knowledge and think critically? Think about your own life. Sure, your education played a part in your success, but other factors like how you present yourself, your ethics, and your drive were just as important. We all know that the stuff we memorize for exams is far from our brains. Yet, we're still making students do the same thing, loading them up with debt, and then they struggle to find jobs to pay it off. By the time they're debt-free, they're forty-five and wondering where their youth went.

As educators and investors, we need to rethink how we prepare the next generation. Shouldn't we focus on project-based learning? What if students work on building a bridge instead of cramming for exams? They learn the principles, collaborate, and apply their knowledge. This hands-on approach should be the norm from elementary school to college.

The point of education is to succeed in real life, contributing to society in meaningful ways. It's not about collecting fancy degrees and then feeling lost and jobless. Many people are now questioning the value of traditional higher education.[69] At this junction of technology and education, you have the power to challenge the old methods. Let's move away from rote memorization and focus on critical thinking and practical application. Give students the tools they need, including AI, and watch them thrive.

Education shouldn't be about prestigious certificates but factual learning and contribution. It's time to shake things up, question the status quo, and be brave enough to change the world. It isn't easy, but our kids and future are worth it. As investors, this is your chance to make history.

Take education matter in your own hand

When choosing future schools or colleges for your children, you should focus on institutions pioneering AI integration within their curriculum. Look for schools teaching coding, machine learning, and data science early on, whether in middle school, high school, or higher education. These subjects should be part of hands-on projects, offering real-world AI tools and fostering interdisciplinary learning.

Additionally, seek out schools that create an environment of innovation, critical thinking, and problem-solving, all essential skills for thriving in an AI-driven world. Institutions collaborating with tech companies to provide students with practical AI applications stand out. Ensuring your children are exposed to these elements will equip them with the skills and mindset needed to navigate and excel in the future landscape of education and work. This comprehensive approach will prepare them for a world where AI is an integral part of everyday life, ensuring they not only keep up with but lead in technological advancements. We are hopeful that with the concerted collaboration of government, industry, and the public sector on these issues, the continued advances in AI will come to be a powerful aid to more equitable and just educational systems and an ingredient to engaging, innovative learning environments that will serve the needs of all our diverse learners and educators.

AI has the potential to create engaging, innovative learning environments that cater to the needs of all our diverse learners and educators, making education more equitable and just. Embrace the journey of integrating AI into education. Let's work together to build a future where technology empowers educators and students alike, fostering a brighter and more inclusive world of learning for the next generations of humAIns.

As you have developed a deep understanding of AI and its applications in education, grasping how to navigate a polarized world is imperative. How do we sift through the noise in an environment where good and evil, truth and lies, and stark contrasts such as red and blue or black and white coexist? How do we maintain our integrity and continue to advocate for what we believe to be true amidst the age of misinformation? The next chapter will explore how AI can generate and combat misinformation and discuss strategies to protect ourselves, our families, and our society.

Challenge your thinking

- Problem-based learning (PBL) is a different way of teaching where students learn by tackling real-world situations. Instead of just memorizing facts, they work together to find solutions. This approach focuses on the "how" of learning, encouraging critical thinking and problem-solving skills rather than just focusing on the "what" or the information itself. Some people believe that if we change our entire education system to PBL, students will be better prepared for the challenges they face in the real world.

- With all the fantastic AI tools available, maybe it's time to encourage students to use them to think creatively and solve problems in new ways. Instead of banning AI, we should embrace it as a learning tool. If we don't, we might be holding students back from developing essential skills for the future.

chapter 6

MISINFORMATION FREE FUTURE

Enlightenment is man's emergence from his self-incurred immaturity. - Immanuel Kant

Eternal happiness for the turkey

Once upon a time, there was a turkey named Tom. Tom lived a typical turkey life, foraging for food and competing with his five turkey siblings under the watchful eye of Farmer Joe. It wasn't an easy life, but Tom was getting by just fine, even if he had to hustle for his meals.

One summer, something peculiar happened. Farmer Joe started giving Tom extra food—

more than his siblings received. Tom was over the moon. "I must be the best turkey ever!" he thought. He strutted around, flaunting his feathers, convinced that his hard work and dazzling beauty had won the farmer's favor. Day after day, Tom gobbled up the extra food, getting fatter and happier. He didn't question it. Why would he? Life was suddenly fantastic. Extra food meant extra dopamine, and Tom was living the turkey dream.

Tom noticed some changes as the days turned into weeks and the weeks into months. The leaves started to fall, and there was a chill in the air. November had arrived, but Tom didn't think much of it. One crisp morning, Farmer Joe and some friends came to the pen. They grabbed Tom and carried him away. "Wow, this must be something special," Tom thought as they laid him on a chopping block. His thoughts of grandeur lasted until—whack—his head fell off. Tom never saw it coming.

Tom-the-Turkey's story is a classic example of cognitive anchoring. He anchored onto the belief that the extra food meant he was special, ignoring any contrary signs. Cognitive anchoring occurs when we latch onto specific information and ignore opposing evidence, creating a polarized view. Tom's growing food supply seemed like a blessing, making him ignore the change of seasons and the impending Thanksgiving. This one-sided perspective kept Tom in blissful ignorance until the end. Like Tom, people often experience unexpected tragedies because they don't seek or recognize diverse information. One of the main reasons is misinformation and disinformation are so intricately integrated within our information environment that it is extremely difficult to distinguish clues from patterns. In Tom's case, he missed the clues—like the change in seasons—that hinted at his fate. This tale humorously, yet poignantly, reminds us to look beyond our immediate perceptions and consider multiple viewpoints to avoid being blindsided by reality.

This chapter will explore the profound political and ideological divide within our country and globally resulting from constricted viewpoints. We will examine how misinformation is critical in exacerbating this divide, often serving the interests of a select few who gain financially from the discord. Following this, we will explore the concepts of biases and heuristics, revealing how these cognitive shortcuts can lead to flawed decision-making. Finally, you will learn practical and effective strategies for leveraging AI to counteract these biases, enabling you to make more informed decisions and enhance your life skills in the fight against misinformation.

The deep divide
A nation divided by the numbers

The modern attention span lasts between 3 seconds to 30 minutes, depending on the media page format. Some social media platforms limit video or music lengths, and if your message can't get through in under 60 seconds, your content won't trend. Trying to make sense of the explosive division in politics, climate issues, and racial polarization isn't as easy as making vegetable soup. Over the last decade, polarization in the U.S. has been worse than at any other time in American history. According to a 2021 Pew Research Center study, about 42% of Republicans and 41% of Democrats view members of the opposing party as more immoral compared to each other.[70]

This divide extends beyond public discourse and personal lives, shaping our most intimate relationships. The American Enterprise Institute reports that 15% percent of Americans would be unhappy if a close family member married someone from the opposite political party. For Democrats, this figure rises to about 20 percent, and for Republicans, it's around 16 percent.[71] These numbers reveal that political affiliation has become a fundamental part of our identity, influencing even our closest relationships. Think about it when being at a family gathering and knowing that one in five people there would be upset if you married someone with different political views. It's a stark reminder of how deeply entrenched these divisions have become and their impact on our lives. The divide of polarization extends beyond the USA.

Global polarization: The Brexit bombshell

Let's fast-forward to 2010 following the Labour government maintaining power throughout the 1990s and 2000s when the UK's involvement in the Iraq War had to start paying its bills. The financial crisis that hit the US hard 2008 also slammed the UK. Following serious hardships, the country decided to vote to either leave the EU or remain a member. The elected Conservative Prime Minister promised a referendum—spurred by the sudden appearance of a third political party, the UK Independence Party (UKIP), which threw sand in the clockwork of the government by running on anti-politics and anti-government choices. The ballet was binary: Leave the EU/Remain in the EU. By 2016, when a significant political sway happened in the US, the Brexit initiative occurred in the UK. It was considered one of British history's strangest and most influential political events. The short story suggests Brexit was a means to an end for everyone to get ahead financially. However, the prime minister (2010 to 2016) dismantled the political agenda, hoping to defeat the anti-EU faction in the Conservative Party using Brexit as its foundation.[72]

But what does all this history have to do with AI? The digital age has revealed the fragility of democracy. Social media, a powerful tool in the hands of the public, was weaponized to manipulate public opinion and influence voters to support the Brexit initiative.[73] The use of AI-driven targeted advertising, coupled with the tactics of the political consulting firm Cambridge Analytica, led to a proliferation of disinformation and propaganda. These tactics stifled rational discourse and objective analysis, with synthetic personas masquerading as real people, easily swaying the attention-defeated masses and sowing discord within the electorate.

A world in flux

The AI-drive manipulation isn't limited to Europe. In Brazil, the election of Jair Bolsonaro (38th president of Brazil from 2019 to 2023), a far-right candidate, brought deep societal divides over corruption, crime, and economic policy to the forefront. Bolsonaro's presidency has been marked by significant controversy and opposition from left-leaning groups, showcasing the sharp ideological splits within the country. In the Philippines, President Rodrigo Duterte's (2016 to 2022) tough-on-crime approach and his controversial war on drugs have polarized the nation.[74] Duterte's policies are supported by those who seek a hardline stance on crime but are vehemently opposed by human rights advocates and left-leaning political groups.

Chatbots already openly answer public questions. Data collection and analysis have transformed government services. Digital information and machine learning plan applications and manage supply changes. Brexit and political agendas sparked by AI misuse remind us of how we're navigating uncharted digital pathways. The unchecked power of AI and social media has compromised the foundations of legitimate democracy, underscoring the urgent need for regulation and oversight in the digital age. You must remain proactive, diligent, and unwavering in your commitment to preserving AI integrity.

The role of social media AI algorithms

Platforms like Facebook, X (formerly Twitter), TikTok, and Instagram have revolutionized how we communicate and access instant information. They've also played significant roles in deepening political divides. However, it's crucial to remember that these tools also have the potential to unite people for positive change. The Arab Spring is a powerful example, where social media mobilized protests and political change, creating "a series of pro-democracy uprisings that enveloped several largely Muslim countries, including Tunisia, Morocco, Syria, Libya, Egypt, and Bahrain." This inspiring event demonstrated the power of media platforms to facilitate social movements and democratize information, instilling hope for a more connected and informed future.

Conversely, the same tools that can unite people for a common cause can also create echo chambers, where users are only exposed to information that reinforces their existing beliefs. A study by the Pew Research Center found that about 64 percent of Americans believe that fake news confuses basic facts of current events.[75] Many influencers out there are willing to do everything and anything to get your attention, capture a "like," and start a trend that might not benefit anyone except the person running the media group. This issue is not just confined to the United States; it's a global concern, a shared challenge that connects us all.

Social media platforms are designed to keep us engaged by showing us content that aligns with our views and *likes*. This creates an immediate feedback loop, where continent exposure to information reinforces our beliefs and biases. Advertisers have AI spies watching for keyword searches to drop product ads into your accounts and even send advertisements to your TV feeds. A 2018 study published in the Proceedings of the National Academy of Sciences found that "Social media may limit the exposure to diverse perspectives and favor the formation of groups of like-minded users framing and reinforcing a shared narrative, that is, echo chambers."[76] You don't want to be the Turkey of our story.

Most people scrolling through your Facebook feed only see posts from friends and pages that share their political views. If you 'friend' someone and don't follow their feeds, 'like' their posts, their pages drop into obscurity where your great aunt and weird cousins live in the digital weeds because you're not engaged in their media posts. Over time, certain groups and persons you interact with more often reinforce your beliefs and make understanding or tolerating opposing viewpoints harder. This phenomenon is known as *"confirmation bias,"* where you seek information that confirms your preexisting beliefs and ignores information that contradicts everything you once believed.[77] That's why the Flat Earth Society continues to thrive, and 23-year-old Peter McIndoe's viral *Birds Aren't Real* movement started with memes and trending.

Hackers and AI: A new frontier of misinformation

Deepfakes have evolved with technological progress. Free software, accessible to anyone, can transform your favorite actor into a new persona.[78] The widespread use of synthetic media powered by AI blurs the line between reality and fiction, generating images, realistic video footage, and texts of events that never occurred. Unfortunately, deepfakes are becoming increasingly difficult to detect in a world of fleeting attention spans. As visual creatures, humans tend to trust what we see immediately. The Generative Adversarial Network (GAN) is designed to learn and enhance characteristics, enabling it to create state-of-the-art digital manipulations of audiovisual scenes that can deceive most people. The technology behind these potentially harmful manipulations is inexpensive and requires basic AI skills, training, and programming that isn't resource-intensive, making it easily accessible to a wide audience.

The 2016 US presidential elections revealed a shocking reality: Deepfakes and AI technology had advanced significantly, unnoticed. After the dust settled from the unusual election campaigns, what truly stunned Americans was the extensive infiltration of hacker groups into the American mainstream media, voter databases, and political campaigns. The use of Xr, Facebook, Instagram, and YouTube for widespread propaganda was the fuse that threatened to implode American democracy. The US intelligence community was inundated with terabytes of misinformation, a testament to the scale of foreign influence efforts.[79]

The hackers used digital guerilla warfare, tapping youngster hackers to blanket social media platforms with propaganda. The efforts of not-so-subtle influencers developed 'troll farms' with AI technology that sent disparaging messages and spread false narratives across social media. Many foreign hackers still face indictments and criminal charges in the US, including conspiracy to defraud the United States, conspiracy to commit bank fraud, and identity theft.

There are no facts, only interpretations

We fall in line with short attention spans and take mental short-cuts in deciding basic human needs and beliefs. You have an expectation of how someone acts given a particular profession. If you go to a doctor's office, you expect to see someone knowledgeable, in specific attire, judging them on stereotypical ideas about doctors. So, you wouldn't expect someone to wander into the patient room wearing speedos, and if they did, you'd expect them to at least wear sandals and a white lab coat. The notion of expectations is a form of judgmental heuristics. Each of us carries a laundry list of heuristic biases.

Since you removed critical thinking from your online doom scrolling, you're prone to believe what your eyes see before your brain uncovers the details. Remember that gorilla in the room? AI can curb your thinking by catering to self-control, concentration, and focus. Impulsiveness can kill quality time. Your heuristics or mental shortcuts facilitate decision-making and problem-solving, generalizing and reducing cognitive load. Without getting too bogged down in something you'll gloss over, let's use a few examples and move on. At least you'll have enough talking points for anecdotes over coffee with someone close to you.

Traveling by vehicle feels safer than air travel because you have no trouble imagining a plane load of people careening out of the sky in a flaming mess of twisted metal. You base your travel arrangements based on that availability heuristic. Or if you're worried about someone on the street at night wearing a hoodie, you have a notion they're out to get you, not just out and about, walking with the hood up because they saw you walking in the dark alley and wanted to avoid eye contact. Your brain uses heuristics to evaluate situations without deliberate thought processes.

Why does this matter with AI? The influence driving you to make decisions based on certain biases you see and the heuristic effects around those ideas are geared toward your notions about AI technology. Throughout this text, you've kept the concept of AI global domination. Was that a bias ahead of your reading? We all probably have it because our imaginations are far greater than any AI tech. Ever heard of the uncanny valley effect? In the

1970s, Masahiro Mori proposed that the uncanny valley was a "phenomenon wherein humanoid artificial characters may some-times cause uneasiness in human beings."[80] AI is the root of this generalization. You may be thinking that getting vetted by my online group can help me see the real truth. Let's look into it.

Maybe group human judgment is better – Not So

Groupthink is a psychological phenomenon where the desire for harmony or conformity in a group results in irrational or dysfunctional decision-making. It's like being on a bus headed for a cliff, but no one speaks up because they don't want to rock the boat. Groupthink can be dangerous, leading to poor decisions that have far-reaching and potentially disastrous consequences. One of the most striking examples of groupthink occurred during the lead-up to the 2008 financial crisis. Financial institutions worldwide fell into the trap of believing that the housing market was stable and risky loans were safe. The collective delusion was fueled by the reinforcing nature of groupthink, where no one wanted to challenge the prevailing optimistic viewpoint. The result was a massive economic downturn that affected millions of people globally. The event shows how groupthink can amplify individual biases, leading to disastrous outcomes.

Another historical example is the Bay of Pigs invasion in 1961. President John F. Kennedy and his advisers were so eager to agree and avoid conflict that they failed to analyze their plan to overthrow Fidel Castro in Cuba critically. The mission failed, embarrassing the United States and strengthening Castro's position. This incident highlights how a cohesive group can make catastrophic mistakes when not encouraged to embrace dissenting opinions.

So, why does groupthink happen? It often stems from strong leaders discouraging dissent, creating an environment stifling opposing ideas. In cohesive groups, members may fear that voicing disagreement will disrupt the harmony or jeopardize their position. This fear of conflict can prevent meaningful discussions and lead to a collective tunnel vision. Fostering an environment where diverse perspectives are encouraged and valued is essential to combat groupthink. For instance, President Kennedy learned from the Bay of Pigs debacle and later ensured that his advisers critically evaluated different viewpoints during the Cuban Missile Crisis, leading to a successful resolution.

In modern corporate settings, some companies have implemented strategies to prevent groupthink. For example, at Bridgewater Associates, a leading investment firm, employees are encouraged to openly challenge each other's ideas. This radical transparency culture helps uncover blind spots and make more informed decisions. By being aware of the dangers of groupthink and actively promoting critical thinking and dissent, we can make better individual and collective decisions. Each individual's voice is crucial in this process. Encouraging diverse viewpoints and challenging assumptions are crucial steps in avoiding the pitfalls of groupthink and ensuring that we navigate complex issues effectively. So next time you're in a group decision-making situation, remember to speak up and welcome different perspectives—your voice might be the one that prevents costly mistakes. Now, let us learn how you can use your common sense to fight against misinformation and some lifelong techniques that can make your internal AI smarter.

Commonsense against the misinformation

From birth, you're taught not to believe everything you see and hear. That's why your imaginary friend left your house when you discovered video games and digital media. Exploring the consequences of spreading misinformation online might be an excellent way to teach children and students how to corroborate and check credibility before they believe the earth is flat or birds are hoaxes. Generative AI isn't going extinct anytime soon. It's an integral part of the digital landscape. Children will become early adopters and influencers. They'll need to grapple with the consequences, and someone will need to lead the resistance. Young people on social media already see the negative impacts and privacy issues. Intellectual property (IP) theft is still rampant, and AI generates terabytes of misinformation and disinformation every minute of every day. And unfortunately, young, uneducated people are more likely to believe disinformation generated by AI than misinformation written by humans.[81,82]

Consider the experiment in Chapter 3 and how only 30 percent of the participants noticed the Red Cross or the gorilla. Now, consider that fewer than 3 percent of the participants in a new study spotted false tweets generated by AI over those written by humans. The potential benefits of AI can be easily undermined by AI algorithm biases, highlighting the need for ethical AI development. Until then, your only backup is you. In the next section, you will learn critical thinking skills that can make a pro about spotting fake news. You don't want to be Tom, the Turkey in our story, by waiting till the last moment to understand facts. By then, it will be too late. Let's learn how to fight against cognitive anchoring and focus on pieces of information that can save our existence.

Outsmart the fake news frenzy: Your ultimate guide to becoming a misinformation spotter

In today's lightning-fast digital age, fake news spreads like wild-fire across social media, fueled by clever algorithms that cater to our biases.[83] It's a chaotic landscape where truth and fiction blur. But fear not; you possess the power to rise above the noise and spot the fake news. In this section, you will learn that triangulating opposite information, verifying sources and understanding the motivation for the source can prepare you to combat any misinformation.

Let's assume a scenario: You're scrolling through Facebook and stumble upon a shocking headline claiming, "*Chocolate cures COVID-19*". Thousands of shares? It must be true, right? Wrong! This is your first clue to a phenomenon called cognitive anchoring. The more we see a piece of information, the more we believe it, regardless of its accuracy. Misinformation thrives on trickery. Fake accounts flood the internet with the same false claim, creating a deceptive illusion of credibility known as the false consensus effect. The goal is to make you think, "Everyone's saying it, so it must be true!" Now, how do you fight against it? It's time to get ready for your secret mental weapon.

Your secret weapon: Triangulating opposite information

Let's revisit the chocolate cure. Instead of blindly accepting the claim, unleash the power of search engines and search for the "opposite: "*Chocolate does not cure COVID-19.*" This simple act opens a world of reliable sources that expose the truth. As scientists, we call this null hypothesis testing. Unfortunately, the time has come when you also get to be the scientist in your head, as misinformation is rampant. Without knowing what is true or false, life will be very difficult to live for you and your loved ones. Therefore, triangulating the opposite information is the first step in spotting fake news. As scientists, this is what we do to prove a point. We always search for opposite evidence and finally, if that evidence is not good enough, we conclude with our desired hypothesis. Unfortunately, everyone needs to have that scientist brain to survive in today's digital age to live a fulfilled life.

The very next step is to verify sources.

Verify source: decoding the primary vs. secondary sources

Why Primary Sources Matter

Seek out primary sources like major news outlets and peer-reviewed scientific journals. These sources are like the rock stars of information—thoroughly vetted and backed by evidence. Blogs and secondary articles? They're the backup dancers—entertaining but not always reliable.

Primary sources are the original history manuscripts, providing raw, unfiltered information straight from the source. These are the documents, data, or accounts created by people directly involved in an event or research. They include research papers, official documents like birth certificates or legal contracts, and firsthand accounts such as diaries, interviews, or eyewitness reports. Their reliability lies in the fact that they provide the most direct connection to the facts.

Primary sources are the gateway to unadulterated truth, not interpreted, summarized, or filtered through someone else's perspective. For instance, if you're researching the impacts of climate change, reading the research studies published in scientific journals will give you the data and conclusions drawn by the researchers without any added spin. Suppose you're investigating the causes of a historical event like the fall of the Berlin Wall. Primary sources could include speeches made by political leaders at the time, news footage, photographs, and interviews with people there. These sources give you a front-row seat to the events unfolding and empower you to form your own understanding based on direct evidence.

The role of secondary sources

Secondary sources interpret, analyze, and summarize primary information. They are one step removed from the original data and often provide context, commentary, and analysis. Examples include textbooks, articles in newspapers or magazines, and documentaries. Secondary sources are beneficial for gaining an overview or understanding of a topic's broader implications. However, it's important to approach them cautiously, as they come with the author's interpretations and potential biases; accentuating potential biases can make you feel more cautious and critical in your research.

If you read a history textbook about the fall of the Berlin Wall, the author might summarize events, context, and analysis of its significance. While this is useful for a comprehensive view, it's still important to refer back to the primary sources to see the raw data and firsthand accounts. It emphasizes the importance of primary sources, which can make you feel more responsible and diligent in your research.

Why you should go back to the source

Going back to primary sources ensures you get the whole picture. Secondary sources, while valuable, can sometimes misinterpret or selectively present information based on the author's perspective or agenda. By examining the primary sources, you can verify the accuracy of secondary accounts and avoid being misled by potential biases. The caution and discernment in your research process are crucial for ensuring the accuracy of your information.

Consider the case of the 2008 financial crisis. Primary sources such as the actual mortgage agreements, financial reports, and testimonies from banking executives provide a direct insight into what happened. Secondary sources like documentaries or news articles might highlight specific aspects or interpretations of the crisis, but only by reviewing the primary documents can you fully understand the complexities involved. The satisfaction and sense of accomplishment from fully understanding such a complex issue through primary sources is unparalleled.

Bringing it all together

When conducting research or trying to understand a complex issue, it's essential to use both primary and secondary sources. Primary sources offer the raw data and firsthand accounts that form the foundation of knowledge. Secondary sources provide analysis, context, and interpretation that help you understand the broader implications.

Adopting this approach ensures you are well-informed and can critically assess the information you encounter. It's best to start with primary sources, using them to ground your understanding of indirect evidence. Then, by consulting secondary sources, you can see how others have interpreted that evidence and understand different viewpoints.

Think if you're considering buying a new gadget. The primary source here is the manufacturer's user manual or technical specifications, while secondary sources include reviews from tech bloggers or consumer reports. Reading the primary source gives you a clear understanding of the gadget's capabilities, while secondary sources provide opinions on its real-world performance. However, if you rely solely on secondary sources, you might miss out on crucial information. This is a common pitfall in decision-making, whether about your career, life partner, or other significant life choices.

After you verify the source, the next and final step is to understand the motivation behind the information dissemination.

Uncovering the motivations of the source

Empower yourself by understanding the motivations behind the information you encounter. In an age where misinformation spreads rapidly, knowledge is power. While verifying the source might seem complicated, it becomes more evident when you follow one fundamental principle: follow the money. The core of many motivations, whether in media, advertising, or social content, often boils down to finance and economics. It takes money to propagate information, and those who do it do it because they want a higher investment return.

One morning, you came across an article claiming that chocolate cures COVID-19. At first glance, this might seem like groundbreaking news. However, before you share it, consider the potential motivations behind this claim. Who benefits if people believe this information? If the article leads to increased chocolate sales, the primary beneficiaries would be chocolate companies. To uncover the truth, you might need to investigate who funded the studies or sources cited in the article. Are they financed by chocolate manufacturers or individuals with stakes in the chocolate industry?

Take a moment to consider who stands to gain financially from the spread of certain information. Running ads or disseminating information on social media requires an investment. If someone is willing to spend money to ensure you see this information, they are likely expecting a return on that investment. This expectation can often translate into biased or misleading information designed to influence your behavior for their benefit. Be cautious and discerning in your consumption of such information.

Your toolkit for a misinformation free future

Many experts fear uncivil and manipulative behaviors on the internet will persist – and may get worse, leading to a splintering of social media into AI-patrolled and regulated 'safe spaces' separated from free-for-all zones. Some worry this will hurt the open exchange of ideas and compromise privacy. Doom-scrolling takes on a whole new meaning when considering a bleak future for the AI-dominated digital domain. The global ecosystem of social interaction depends on instant access to *everything*. Your life belongs to the network. Status updates, political advocacy, comment chains, rankings and ratings, omnipresent reviews, and news feeds shape everything you touch online.

Evidence suggests social media spreads extremist causes, weaponized by terrorists, using AI to create better and increasingly effective kinds of propaganda. Respected scholars show proof that social bots disrupted political events and presidential elections. Foreign troll farms slammed the US social media landscapes with fake news.

Did you know more than 62 percent of Americans get their news from social media? Some experts believe we're losing the internet to cultures of hate. Former social media employees testify their employers suppressed conservative news content. Governments are actively increasing their efforts to monitor user social media and instant messaging accounts. Congress and the CIA are actively examining how the Internet will shape the next war. "The weaponized narrative seeks to undermine an opponent's civilization, identity, and will by generating complexity, confusion, and political and social schisms."[84] An analysis published in Scientific American stated AI tools influence us more than we ever realized.

"We are being remotely controlled ever more successfully in this manner. ... The trend goes from programming computers to programming people ... a sort of digital scepter that allows one to govern the masses efficiently without having to involve citizens in democratic processes."

Is there hope for the future with AI ruling the world? Absolutely, and it begins with you. Digital commentary originates from individuals manipulating the system. We're already looking towards the future because there is a widespread demand for technological systems or solutions that promote more inclusive online interactions. Trolling, harassment, and distrust are tiresome. Negative activities require active members, and people grow weary of battles in the online social climate. Networked communication technologies empower you. Your actions can make a difference. You can create much better security and moderation solutions by using AI to identify predictable online reputation systems. This will make it challenging for bad actors to disrupt the system. Purging troll bots will take time, but it's worth a few seconds more to sift through the nonsense to see the bigger picture. However, it's essential to remember that the responsible use of AI and technology is key in combating negative online behavior. Awareness and caution are necessary in this digital age.

Let's recap this necessary skill one last time.

Triangulating Opposite Information: RECAP

When you encounter information online, the first step is to triangulate opposite information by proving it may be misinformation. For instance, if you read that a particular diet plan can cure a disease, don't just take one website's word for it. Find opposite information online first to check its validity. Be aware that there are misinformation AI bots that are programmed to anticipate your search for similar information and manipulate your understanding. They do this by flooding the internet with false information that supports the initial claim, making it harder for you to discern the truth.

Next, always verify the credibility of your sources. Major news outlets, academic institutions, and well-known experts are usually reliable. Look for peer-reviewed studies, reputable news articles, and expert opinions. This practice ensures you're not falling for misinformation from a single, possibly biased source. It's important to be cautious and thorough. Check multiple sources to see if the information holds up. For example, if you come across a groundbreaking health tip, see if it's endorsed by recognized medical journals or health organizations. Trustworthy sources have a reputation to maintain and are less likely to spread false information.

Finally, it's vital to question the motives behind the information source. Ask yourself why it's being presented and who stands to gain. Is there an agenda? Often, financial incentives drive misinformation. If a piece of news claims that a certain product is a miracle cure, investigate who benefits from this claim. Is it a company that stands to gain financially? Understanding these motivations can help you see through potential biases and make more informed decisions, making you more vigilant and aware.

The future of AI integration with human judgment is necessary for bridging the divide and distance we are creating within our families, friends, country and the world. You can combat misinformation and create a healthier digital space by learning to manage and tweak your algorithms. This empowerment means you won't just be a passive consumer of information but an active participant in shaping your AI in the future and emerge as an effective humAIn.

Decision-making falls short when you're constantly bombarded with misinformation, manipulated images on social media, and increasing social isolation. The next chapter will explore these pressing issues, examining their effects on mental health and overall well-being. By learning new skills and fostering a joyful mindset in the age of AI, you can create a happier society for yourself and those around you and realize that you are not alone in facing these challenges.

Challenge your thinking

- Understanding and gaining control over the algorithms that shape our social media feeds is not just a matter of personal preference; it's a fundamental right. These algorithms influence what we see, who we interact with, and how we perceive the world. They can amplify misinformation, create echo chambers, and even impact our mental well-being. The time to demand transparency and control over these powerful tools is now. This isn't just about you; it's about protecting your loved ones, community, and future generations. Let's raise our voices together and demand a say in how AI algorithms shape our lives.

- Constantly fact-checking every piece of information can be mentally exhausting. It's essential to find a balance that works for you. Instead of doubting everything, focus on the information that matters most. Ask yourself: What information will likely have the most significant impact on my life and decisions? (For example, news related to health, finances, politics) What information is most susceptible to manipulation or bias? (For example, social media posts, opinion articles, advertising). You can protect yourself from misinformation without burning out by prioritizing your skepticism and directing your energy toward the most critical information.

part III

AUGMENTATION

chapter 7

THE POSITIVE HUMAIN MINDSET

Only in the eyes of another human being can I find myself fully reflected. - Johann Wolfgang von Goethe

The tale of king Midas: A lesson in avarice and regret

Once upon a time, a wealthy and powerful king named Midas ruled the ancient kingdom of Phrygia. Known far and wide for his immense riches, Midas's insatiable desire for more overshadowed his heart. One day, the god Dionysus visited Midas's realm and offered the king a wish in gratitude for a kind act. Blinded by greed, Midas wished that everything he touched would turn to gold, believing this would bring him unparalleled wealth and happiness.

Initially, Midas was ecstatic. Everything he touched—goblets, furniture, flowers—turned brilliant gold. He imagined himself as the richest man to live, reveling in his newfound power. But soon, joy turned to despair. When Midas tried to eat, every morsel turned to gold, leaving him hungry and desperate. The nightmare worsened when his beloved daughter, trying to comfort him, was turned into a lifeless golden statue at his touch.

Haunted by his folly, Midas prayed fervently to Dionysus, begging him to reverse the curse. Moved by Midas's genuine repentance, Dionysus took pity on him, instructing Midas to wash in the river Pactolus to cleanse himself. The curse lifted as Midas immersed in the waters, and his daughter returned to life. From that day forward, Midas renounced his greed, cherishing the true treasures of life—love, family, and the happiness of his people.

In today's digital age, social media promises the golden touch, offering instant connection, validation, and entertainment. Like Midas, many of us are drawn to the allure of more—more likes, more followers, more content. But this endless pursuit can come at a cost. Just as Midas's golden touch turned his blessings into burdens, the algorithms driving social media can transform a source of joy into one of stress and anxiety.

Social media platforms are designed to maximize user engagement, often by prioritizing sensational or emotionally charged content. This can create a cycle of addiction, where users continuously seek the next hit of validation. However, this engagement can lead to negative emotions, such as jealousy, inadequacy, and loneliness, mirroring Midas's despair when his touch turns his loved ones into gold.

But there is hope. Just as Midas sought forgiveness and transformed his life, we too can approach social media with mindfulness and wisdom. By understanding and taking control of the algorithms that shape our digital experiences, we can break free from the cycle of negativity. Your interactions with AI algorithms can enable you to customize your feeds, promoting content that uplifts and enriches rather than divides and depresses.

In this chapter, you will dive into the broader context of the mental health crisis, shedding light on the alarming prevalence of depression. You'll uncover how profit-driven algorithms exploit the dopamine rush in our brains, leading to addictive behaviors and negative mental health impacts. Finally, you'll discover practical techniques to leverage mindfulness and AI, helping you and your loved ones safeguard your well-being and thrive in today's digital age.

Declining mental health pandemic

Despite the abundance and wealth in Western countries, mental health issues are escalating at an alarming rate. The overall decline in mental health is affecting teens, adults, and older adults alike, with severe implications for society as a whole.

According to the National Institute of Mental Health, about 13.3 percent of adolescents aged 12-17 in the United States experienced at least one major depressive episode in 2017. What's even more alarming is the sharp increase in teen depression rates—rising by nearly 63 percent between 2009 and 2017, as reported by the American Psychological Association. Currently, suicide is the second-leading cause of death for teens and young adults ages 10-34. The CDC reports 10 percent of high school students attempted suicide in the past year (2023). This percentage is highest among females (13%), American Indians/Alaska

Natives (16 percent), black teens (14 percent), and lesbian, gay, or bisexual teens (22 percent).[86]

The World Health Organization (WHO) estimates more than 264 million people worldwide suffer from depression, with a global prevalence of 4.4 percent. Between 2005 and 2015, the number of people living with depression increased by 18.4 percent.[87] This trend has significant implications for worldwide productivity and healthcare systems. The economic impact of depression and anxiety is staggering, costing the global economy $1 trillion each year.[88] This pervasive unhappiness affects personal well-being and shapes worldviews, influencing socio-political events and fostering polarization.

Depression among older adults is a concerning yet often over-looked issue. As populations age, the mental health of older adults becomes increasingly important. According to the CDC, approx-imately 7 million American adults aged 65 and older experience some form of depression. The causes include biological changes, medication side effects, and significant life changes like retire-ment. However, it's important to understand that social factors, such as divorce, living apart from family, and losing friends, can exacerbate feelings of isolation. Divorce rates among adults aged 50 and older have doubled since the 1990s, adding emotional and financial strain that contributes to depression.

The impact of declining mental health is profound. Depres-sion, the leading cause of disability worldwide, is a challenge that requires your attention. In the U.S., roughly 20 percent of adults experience some form of mental illness each year. As of 2023, about 83 percent of Americans are active on social media.[89] While these platforms connect us, they also bring various social and mental health challenges. You should be aware of these issues and take steps to address them. Studies have found a troubling link between high social media use and increased depression symptoms in teens. Social media platforms and AI-driven algorithms, meant to bring joy and connection, can sometimes do the opposite, but with awareness, you can use them more responsibly.

The self-serving algorithm of social media

Technological progress has merely provided us with more efficient means for going backward. Let's try to dig out of this depressive dialogue but still face the truth about mistrust in social media content. Have you ever heard the saying: "You can fool some of the people all of the time and all of the people some of the time, but you cannot fool all of the people all of the time?" According to the internet, the quote appeared in American newspapers between 1886 and 1889. Some diehard scholars think Abraham Lincoln used it in a speech in 1856; others say he used it in 1858. Yet, some scholars believe the poet John Lydgate (1370-1451) first scribbled the phrase later adopted by Lincoln or a newspaper journalist at the time trying to put a positive spin on events leading up to or following the Civil War. Either way you look at it, the phrase is still relevant today. And more than likely, you've seen a meme with a variation of that quote.

AI can't please everyone all the time. No one really knows

what they want, even if Dionysus, the Greek god of wine and ecstasy, showed up at your dinner party and gave everyone a single wish to keep the drunken sailors happy. This analogy is used to illustrate the unpredictable nature of human desires. You'd end up with someone trying to rule the world. It's all fun and games until the god of insanity shows up and ruins your party.

Exposure to disturbing or triggering content is another concern. Algorithms can inadvertently expose users to distressing material, worsening mental health challenges. And let's not forget FOMO, the "fear of missing out." A study in the *Journal of Social and Clinical Psychology* noted that FOMO was strongly linked to higher social media engagement rather than real-life activities, exacerbating feelings of inadequacy and anxiety.[90] Social media also pressures you to curate a perfect online image, adding to a complex web of mental stressors. Balancing the benefits and drawbacks of these platforms is key.[91] Remember, it's not about quitting social media entirely but using it wisely to enhance your life without letting it overshadow your mental well-being.

The negative nature of social media platforms spawns from the profit driven philosophies at the cost of human compassion and kindness.

The profit driven loop

It's not a coincidence social media is addictive—it's by design. Social media platforms use sophisticated AI algorithms meticulously crafted to keep you engaged, ensuring you spend as much time and money on their sites as possible. These algorithms are the driving force behind the addictive nature of social media, working to drive profits while keeping you hooked. Notice those compelling trends, those engaging memes, and those enticing "likes."

When you sign up for a social media platform, you enter into a transparent agreement where the algorithm immediately begins collecting data about you. It tracks every like, share, and comment, and it tracks the time you spend looking at specific posts. This data forms a detailed profile of your interests, preferences, and behaviors. Using this data, the algorithm personalizes your feed to show you content that aligns with your preferences. For instance, if you frequently 'like" fitness-related posts, your feed will show workout videos, diet tips, and motivational memes. Some aggressive AI will make you feel worthless for eating after eight at night or feel awkward for looking in the mirror. This specifically personalized content keeps you engaged because it resonates with your interests and beliefs. Algorithms measure your engagement with different types of content. They note which posts you interact with the most and which you scroll past. This helps the algorithm refine its predictions about what you like. Each interaction provides feedback, tweaking the algorithm to be more accurate over time. The more you engage with certain types of content, the more the algorithm shows you similar content. The reinforcing cycle in which your feed becomes an echo chamber of your beliefs and interests is exactly what advertising vampires want to know about you.

Did you ever get the feeling someone's listening to your conversations? Sometimes, you might talk to someone about a random service or product you've never tried or wanted to use. It could be an anecdote about something you read and wanted to share with someone else, maybe something weird or random—cat diapers—electric table saws. After a few days, you suddenly see an ad drop into your social media feed or an advertisement appears on your streaming service highlighting the latest and greatest electric table cat diapers. There's a good chance the AI software preloaded on your smartphone will always hear you.

The voice assistant features on your smart device allow the AI to tune in and turn on because you've authorized or maybe not accessed your microphone and video features.[92] It listens to your queries and voice commands. AI algorithms pay attention to phrases and keywords. You can change the permission settings by removing the assistant features. But there's a good chance the other person you are chatting with hasn't disabled their AI assistance or even cares, and the ad algorithms know *where* you are and *who* you are and will pour those ads across your social media no matter what.

The primary goal of these algorithms is to maximize and capitalize on the time you spend on the platform, leading to addictive behavior. The longer you stay, the more ads you see, and the narrower the focus is on products or services. Each ad view translates into revenue for the platform. Some companies spend billions on AI to ensure their services or product gets to the right viewer. With detailed profiles, algorithms build an expanding database about you; advertisers can target you with highly personalized ads. Some companies, remember Cambridge Analytica Ltd., sell your data to other advertising vampires. The targeted approach increases your likelihood of clicking on an ad, boosting the platform's ad revenue. Engaged users like you are more valuable to advertisers. When you interact with content, it signals to advertisers that you're an active user, making the platform lucrative for them to spend their ad dollars. This all loops back to the addiction cycle.

The addiction cycle

AI doesn't just operate on your smart devices. It also has a profound influence on your brain. Every like, comment, or share you receive triggers a small release of dopamine, the brain's reward chemical. The AI's understanding of human psychology allows it to make you feel good. It reinforces the behavior, much like how you might feel after eating your favorite food or receiving a compliment. The algorithm creates an echo chamber by constantly showing you content that aligns with your beliefs. You feel validated and *right*, which is inherently satisfying. The need for this validation keeps you coming back to the platform. In the next chapter, we discuss how dopamine is a driving factor for motivation and how you can combat using digital detox.

The algorithm also plays a significant role in ensuring that the platform is constantly updated with new content. This continuous flow of information, orchestrated by AI, instills a fear of missing out on essential updates, driving you to check the platform frequently. Algorithms often prioritize sensational or emotionally charged content because it generates more engagement.[93] This cycle is designed in your brain to create and improve social connections.

No person is an island

Throughout human history, social connections have been instrumental to your survival and advancement as a species. In early hunter-gatherer societies, social bonds were vital for resource sharing, collective hunting, and protection from predators or rival groups. As you evolved, these social networks expanded and became more complex, laying the foundation for developing language, culture, and, eventually, organized civilizations. The drive to connect has always been deeply ingrained in your biology; neuroscientific research has shown that social interaction activates reward pathways in the brain, releasing chemicals like oxytocin and dopamine, which make you feel good and encourage you to maintain these beneficial social bonds.

The importance of social connection extends beyond mere survival. Your overall health and well-being are deeply intertwined. Numerous studies have linked solid social ties to various positive health outcomes, both mental and physical. For instance, the famous Harvard Study of Adult Development, which has spanned over 80 years, found that the quality of one's relationships was a stronger predictor of long-term health and happiness than any other factor, including diet and exercise.[94]

Loneliness and social isolation, on the other hand, have been found to have serious detrimental effects, equivalent to smoking fifteen cigarettes a day in terms of health impact. These feelings are associated with an increased risk of heart disease, depression, cognitive decline, and even a shorter lifespan. Thus, social connections aren't just a luxury but essential to physical and mental health. However, seeing others can create enjoyment in your brain. That is exactly what the social platform is providing you and the next generation.

Social media's impact on teenager's brain

Social media is a huge part of teenage life. If you were born in the 90s or earlier, there's a good chance you don't know much about social media. However, teens know *everything* about any subject. And social media will reinforce that sentiment. Teen temperament and personalities affected by social media content can displace healthy brain activities. Face-to-face communication is a thing of the past. Sleep is overrated, and physical activities are only as useful as performing trending video segments.

Do you know why kids shouldn't drink alcohol? It's not because it's against the law. It's against the law because teen brains under development are severely injured when introduced to alcoholic beverages too early in life. Adults already know it's not a good idea to jump off the porch roof, using a sheet tied around the neck or an umbrella to catch the fall, because they have already developed a strong braking system in the brain called *executive functioning*. Irrational acting—like throwing yourself on the floor when you are angry—seems like a great idea when you're trying to get out of doing work, but it's not helpful behavior, and adults have better impulse control.

Teenagers' soft and mushy prefrontal cortex isn't fully developed, so they can't rely on *executive functioning*.[94] It's nearly impossible for youngsters to avoid seeking out trouble. And they crave approval from their peers. They know trying to fit a light bulb in their mouths isn't a good idea, but they saw someone on social media do it, so why not try it? The teenage brain wants to be a part of something; it needs to fit in, so swallowing garden slugs might be gross, but it could lure more followers if the kid survives the challenge. What we know, the facts we've learned from former data scientists at monster social media sites, is that social media rots teenagers' brains. Seventeen percent of teen girls said that eating disorders worsened after using Instagram. 32% of teen girls said that they felt terrible about their bodies after using Instagram.[95] The core rationale for still using the social media platform is derived from the desire to connect with others.

Human mind desires connection

Human beings are weird and wired for connection. Close relationships and strong social connections are essential to long-term health and happiness. This intrinsic need for social interaction is why social media has become a significant part of our lives. It's digitized our social connections, allowing us to stay in touch with friends and family across the globe. Social media can enrich your life, but balance is crucial. Use it wisely.

Social media can be a powerful tool for maintaining relationships and sharing information. However, those same algorithms that keep you connected can also trap you in echo chambers, reinforcing biases and increasing polarization. Platforms like

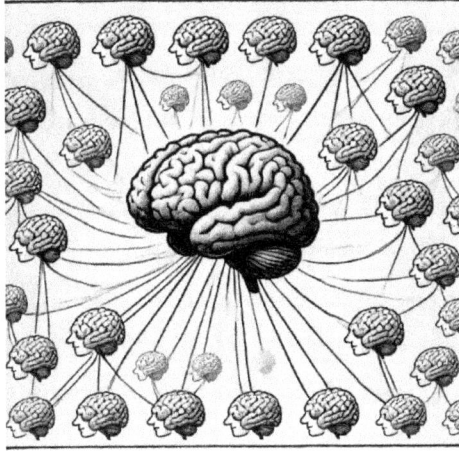

Facebook, TikTok, and Xs often prioritize content designed to provoke strong reactions, which can lead to increased anxiety, depression, and loneliness. According to studies, excessive social media use is linked to these mental health issues, making it essential to approach these platforms with awareness and discernment.[96]

So, remember King Midas and his golden touch next time you log in. Enjoy social media's connections and conveniences, but stay aware of its potential downsides. There may be just the pleasure of connection that drives us to social media.

With or without social media

What is your ambivalence toward social media? Do you recognize its benefits but are wary of its potential harms? It's not about rejecting social media altogether but about finding a way to live with it that enhances our well-being. By being mindful of how you use these platforms and manipulating the AI trolls, you can ensure they serve you rather than distract you and make you buy cheap products from shady discount websites.

Have you ever found yourself endlessly scrolling through your social media feeds, wondering how it's possible to lose track of time so easily? You're not alone. Social media platforms are brilliantly designed to capture your attention and subtly shape your perspectives. If you've seen Netflix's documentary *The Social Dilemma*, you'll know exactly what I'm talking about.[97] The film dives deep into how these platforms use persuasive technology to nurture addiction and influence behavior. They've turned your digital world into a modern-day "Midas Touch," promising connection but often delivering a mixed bag of consequences.

Quitting social media might seem drastic, especially if you were born after 2010 and impractical for most of us. Instead, awareness is your first line of defense. Let's do a mind exercise. Think of a "reality swap" exercise where you and a friend exchange social media feed for a day. You'd see how different information shapes perceptions and could foster a better understanding of differing viewpoints. Remember, if you're not paying for the product, you likely are the product. These platforms are not just about keeping you connected; they have agendas shaped by financial motives.

Why should YOU care

It feels good to scroll and watch clickbait and videos. But there is a hidden cost that you may need to be made aware of. In an age marked by rapid technological advances and an increasingly interconnected world, the need to understand AI is no longer a luxury—it is a necessity, especially when it comes to safeguarding our mental health and the emotional well-being of our children. Social media algorithms seek emotionally dependent people, creating a codependency with AI.

Your salvation to survive the maddening onslaught of AI-driven advertisers and social media echo chambers is to understand how the content controls your attention. The psychological relationship or condition creating a strong desire for approval manifests in an unhealthy attachment to someone or something. You're looking at a codependent relationship with social media. External validation, the need for constant validation, doesn't mean you're a vapid actor on a failed streaming service series; it means you are human and have a rooted sense of self. As a social animal, it's not surprising that you might post something to your feed to gain approval from strangers. Capturing your likeness with the smartphone and posting it for the world doesn't mean the images will steal your soul, but using social media as a mirror isn't healthy.

Social media influencers live in the now; they react to the instant gratification of their likeness being seen, "liked," and shared across countless platforms as validation. Looking for attention from a faceless audience might stimulate you, but it warps your opinion of yourself. Many influencers who hit high traffic, adding to their base and catering to everyone else's needs, lose track of what's important to keep them healthy, sane, and happy. Their nearsighted attention can't focus on the future or see how their influence might negatively impact others. The echo chamber closes the focus, carefully crafting a rampant network and social marketing strategy, keeping viewers engaged as long as the influencers stay relevant.

Social media is unable to respect boundaries. The internet won't tell you "No" and feeds the need to receive online validation. You'll find self-sacrificing time takes away all quality time because you're caught in the dependent cycle to put up a show for others. You fail to see anything beyond how subscribers see you because your codependent relationship with social media keeps you awake at night worrying about others' opinions and needs.

You can't think for yourself because your audience needs to decide for you. Does that top look good with those jeans? Do those jeans go with those shoes? Doesn't that fast-food meal you're about to eat look good on your media feed? It tastes terrible and sheds years off your life while adding inches to your waistline, but don't worry. The fast-food advertisers want *your* opinion about their product.

Guerilla marketing exploits social media influencers using innovative, interactive, interesting methods to sell, sell, sell. Often, it doesn't appear to market any given service or product; it only engages the brain, allowing you to subconsciously remember the unconventional approach. You're on fire, trending, so you'll take that money from the advertisers. They know you have over one million subscribers. They send you scripts to read to your 'followers,' and they can use your likeness anytime they want because you invited the vampires into your social media to feed off you and your masses. Those companies use influencers to advertise their products and services because it's cheaper than paying professional actors, directors, set designers, locations, and scriptwriters. You can sell their goods, get a few more *likes*, gain a few more followers, and forget about putting away any funds for the future because right now, you're more important than anyone else on the planet because your followers say so.

Consider your content, multiple platforms, and how much time you spend interacting with social media. Do you have a codependent relationship with social media? Can you read the signs? Are you addicted to external validation? Do you care about or respect boundaries? Are you too worried about others' opinions? You can't see the natural person in the mirror because you don't have the same AI filters in real life that are on your smart devices.

In a time of deceit, telling the truth is a revolutionary act

Social media companies increasingly prioritize transparency and giving users more control over their feeds. For instance, Instagram has introduced a "Why am I seeing this?" feature that explains why certain content appears in your feed. The component allows you to understand the factors influencing your recommendations. Similarly, TikTok's "For You" page now lets users tailor their feeds by selecting preferences for specific topics or creators. These features empower you to shape your social media experience more actively. Platforms are also releasing algorithmic transparency reports. Facebook, for example, provides detailed reports on how their algorithms function and their impact on user behavior. These reports help researchers and the public gain insights into the complexities of these systems, fostering greater trust and understanding.

Algorithms prioritize original content over re-shares to enhance user experience. Instagram, for example, now favors

Reels created within the platform, promoting fresh perspectives and rewarding content creators. Additionally, algorithms constantly rewrite to encourage meaningful interactions. LinkedIn's algorithm, for instance, prioritizes posts that generate discussions within professional circles, giving more weight to comments and shares than simple likes. Efforts to reduce misinformation and hate speech are also underway. X's algorithm downranks tweets containing misinformation, and Facebook is investing in AI and human moderation to limit the spread of harmful content. These measures aim to create a safer and more reliable online environment. Social media platforms are taking significant steps to address concerns about mental health and data privacy. Instagram has introduced "Take a Break" reminders to encourage users to step away from the app, while TikTok is testing features to limit screen time for younger users. These initiatives aim to mitigate the negative impact of excessive social media use on mental health.

Despite these positive steps, challenges and debates remain. Remember those black-box AI algorithms? It's difficult for the public to understand their work, leading to total mistrust. Additionally, algorithmic changes can have unintended consequences, such as amplifying filter bubbles or inadvertently suppressing certain voices. Governments worldwide are considering regulatory frameworks to hold social media platforms accountable for their algorithmic choices. This evolving landscape highlights the need for ongoing adaptation to balance personalization, control, and societal well-being.

Shareholders are savvy about how their product needs to keep control and stay relevant in an ever-changing environment of shorter attention spans. They lean on programmers and coders to create tools that allow people to seek benefits through social media channels. The only place you feel relevant may be online; it's like that for many teenagers. Isolation and ostracizing aren't always about identity and geographic regions. Currently, social media giants know the harm and damage they cause, so you have instant access to educational content on a vast array of topics.

To find yourself, think for yourself

AI is fundamentally transforming how social media algorithms operate, putting control directly into your hands. AI is changing the world. You won't need to know any programming languages or complex coding; instead, you can tell the algorithm what kind of content you want to see and what you want to avoid. This ability will enable you to create a personalized AI algorithm within your social media platforms, helping you connect with people more meaningfully. However, to make the most of this opportunity, it's crucial to understand the current pitfalls of social media and why these algorithms can sometimes cause harm. Social media algorithms are currently designed to maximize user engagement, often by showing you sensational or emotionally charged content that keeps you hooked. Breaking the cycle of addiction, anxiety, and depression begins and ends with you.

Take a moment to reflect on your social media habits. Are you feeling pangs of jealousy seeing someone's seemingly perfect vacation photos? Or maybe you caught yourself doom scrolling to make you feel inadequate or angry. Social media platforms exploit our psychological vulnerabilities, reinforcing our need for validation and approval through 'likes' and comments. In the future, as you gain the ability to tweak these algorithms, it's essential to practice mindfulness and understand the impact of your choices. If you continue to engage with content that triggers negative emotions, the exact harmful cycle will persist. But if you consciously seek positive, uplifting content and expose yourself to diverse perspectives, you can break free from the echo chamber effect. It doesn't hurt to clear your browser history and clear the history on YouTube. It immediately switches AI to panic mode and requires fresh searches to activate recommendations.

One way to achieve balanced social media is by practicing empathy. When you come across a post that you disagree with, take a moment to consider the other person's viewpoint. Why do they believe what they do? What experiences might have shaped their perspective? You can foster healthier and more constructive dialogues by approaching social media interactions with compassion and an open mind. Sometimes, delete the response before it's published and think how easier it is to swipe off the page than start antagonizing others to dismiss your opinion. Another good option is to practice the technique we learned in the last chapter, "Triangulating Opposite Information." If you have already forgotten, please revisit the recap in the last section of the chapter.

Eventually, highly tailored content and recommendations will enhance the personalization of your daily feed. AI will get hardwired into your brain and help filter irrelevant, low-quality content that doesn't matter to you. Moreover, it's essential to recognize the importance of positively connecting with others. Social media can be a powerful tool for building relationships and community, but only if used thoughtfully. Instead of focusing on the number of likes or followers, prioritize meaningful interactions.

Share content that brings joy, inspiration, or valuable information to others. Engage in discussions that promote understanding and respect rather than conflict and division.

The future of mental health with AI is not just about managing crises but preventing them. AI can predict depressive episodes by analyzing your online activity and alerting you and your healthcare provider before things escalate. It's like having a digital guardian angel looking out for your well-being. As we navigate this rapidly evolving digital landscape, it's crucial to remember the tale of King Midas. Just as he learned to value the true treasures in life, we must approach technology with mindfulness and wisdom. AI has the potential to transform social media from a source of stress into a tool for enhancing our mental health. By embracing this technology responsibly, you can ensure that your digital experience enriches and nurtures your mind and soul.

So, next time you log in, remember that AI is not just about algorithms and data; it's about creating a world where you feel supported, understood, and connected. The future of mental health is in your hands, and it's looking brighter than ever. Embrace it, and let AI help you lead a happier, healthier life and be the best humAIn.

The next chapter explores dopamine—a crucial neurotransmitter that drives motivation, happiness, and daily actions. Together, we'll uncover how dopamine regulates pleasure in our brains. We will discuss how regulating dopamine by digital detox can make you stronger to control your AI instead of being controlled.

Challenge your thinking

- How can you start an influential conversation with your kids or loved ones about the negative biases of profit-driven AI algorithms in social media? Do you need extra training to explain what happens and why it matters?

- We are constantly watching social media and enjoying others' naïve actions. Are you aware of any actions of your own that may seem naïve to you? Remember, we are all products in the eyes of these social media companies. Have you heard of the concept " *Schadenfreude"?* look it up.

chapter 8

DIGITAL DETOX FROM DOPAMINE OVERLOAD

What does not kill you makes you stronger
- Friedrich Nietzsche

The tale of Hameleon's flute

Once upon a time, in a quaint village nestled between rolling hills and lush forests, there lived a mysterious figure named Hameleon. Known for his enchanting flute music, Hameleon had the power to captivate any living creature. His music was so mesmerizing that it could make the stars twinkle brighter and the flowers bloom in rhythmic harmony. However, his most spellbinding effect was on the village rats.

The village was overrun with rats, and despite numerous attempts to control them, the population only seemed to grow.

Hearing of Hameleon's extraordinary gift, the villagers beseeched him to help rid them of the rodent plague. With a nod of agreement, Hameleon lifted his flute to his lips and began to play.

The melody that flowed from Hameleon's flute was unlike anything the villagers had ever heard. It was sweet and haunting, filled with an irresistible allure that tugged at the very souls of those who listened. The rats, one by one, emerged from their hiding places, eyes glazed and bodies swaying in time with the music. They followed Hameleon, entranced, as he walked through the village and out into the wilderness.

The spellbound procession continued until they reached a high cliff overlooking the sea. Without hesitation, Hameleon stepped aside as the rats, still under the spell of the flute, marched straight off the edge and disappeared into the waves below. The village was free, and Hameleon vanished as mysteriously as he had arrived.

Hameleon's flute can be likened to the addictive nature of dopamine in our brains. Dopamine, a neurotransmitter associated with pleasure and reward, plays a significant role in how we experience joy and satisfaction. In moderation, it motivates us to achieve goals and rewards us with a sense of fulfillment. However, in excess, dopamine can lead to addiction and destructive behaviors.

Social media platforms are modern-day flutes, designed to captivate our minds like Hameleon's music entranced the rats. Each notification, like, and comment triggers a dopamine release, creating a cycle of craving and reward that keeps us hooked. Just as the rats followed the enchanting music without questioning, we often scroll through our feeds, oblivious to the passage of time and the potential consequences of our addiction.

We must take conscious steps to manage our social media consumption to break free from this cycle, much like the rats needed to cover their ears to survive Hameleon's spell. Setting boundaries, turning off notifications, and engaging in activities that don't rely on digital validation are crucial strategies. Doing so protects our mental well-being and prevents us from metaphorically marching off a cliff.

In this chapter, we dive deep into the reasons behind our attachment to digital technology—whether it's social media, video games, or blogs. Like the rats entranced by Hameleon's flute, we find ourselves drawn to these digital platforms, often without realizing the extent of our addiction. But here's the crucial part: we'll explore how you can break free from this cycle and avoid metaphorically falling off the cliff. We live in an era where digital addiction subtly yet powerfully controls us using AI. This chapter aims to raise your awareness and encourage you to take that first step towards reclaiming your autonomy. Understanding what makes us susceptible to these digital traps is the first step, and yes, it all boils down to your brain and a specific neurotransmitter called dopamine.

The double edge sword

Dopamine is a neurotransmitter that plays a crucial role in how you feel pleasure. It helps regulate movement, attention, learning, and emotional responses. Dopamine drives us to seek rewards and achieve goals, motivates us to excel, and finds meaning in our lives. When dopamine levels spike, it can lead to dependency and addiction. Most of the time, you don't realize you have a problem until it presents itself in life-changing or awkward moments.

Currently, more than 4.8 billion people actively use social media. It's estimated to expand to nearly 6 billion users by 2027. On average, you spend 2.27 hours daily on social media. And statistically, more than 210 million people have a social media addiction.[98] Social media is engineered to captivate our brains, and teens are particularly vulnerable to its addictive nature. Excessive use of social media can rewire the brains of young children and teens, making them crave immediate gratification. This can result in obsessive, compulsive, and addictive behaviors, which can exacerbate mental health disorders such as anxiety, depression, ADHD, and body dysmorphia.

Social media platforms use AI to exploit your dopamine pathways to keep you hooked. The compulsion mirrors the mechanisms of opioid addiction, constantly triggering dopamine release to keep you engaged and craving more. The influence on developing brains puts very young people in danger of future addictive behaviors, presenting the risks of developing some disorders. It seems unlikely social media is the gateway drug, the same argument happened with movies and video games. Another important area of the brain that gets influenced by dopamine is Pain. Pain, both physical and emotional, can also be influenced by dopamine. In chronic pain conditions, the brain's ability to regulate dopamine can be disrupted, making pain feel more intense and harder to manage. This imbalance can lead to a vicious cycle where pain leads to more stress and anxiety, further throwing off dopamine levels and making it even tougher to find relief.

This chapter is dedicated to understanding the dopamine mechanisms for deciphering digital detox. It's not just about science but also about real-life implications as we embark on our journey with more AI integrated into our daily lives. By understanding how dopamine works, we can better navigate our addiction behaviors, whether it's in the context of substance use or digital consumption, better regulate our perception of pain and take steps towards healthier, more balanced lives utilizing digital detox.

How dopamine regulates our lives
Dopamine and motivation: Your inner drive

The natural release progression of dopamine from nerve cells stimulates different parts of your brain. It can control your movements. For example, lack of dopamine release is linked to Parkinson's disease. Imbalances can cause depression, schizophrenia, and other mental illnesses. When you are working towards a goal, the anticipation of success, the excitement of planning, and the satisfaction of small milestones along the way – all these involve dopamine. It keeps you focused and motivated, providing the mental stamina to push through challenges and setbacks. Whether studying for an exam, training for a marathon, or working on a project, dopamine is your trusty sidekick, cheering you on every step of the way.

The average person touches their face 69 times per hour, but the average of typing, tapping, and swiping handheld devices is around 2,600 times daily. You are intimately intertwined with your digital life even when not using your smartphone. The devices tether us to the rest of the world. You carry around 2 billion potential connections in your device, while as a species; your thriving social structure tends to involve only about 150 individuals.[99] That statistic includes your relations, friends, relatives, and people from your past, and likely won't change in your lifetime. Let's investigate the addiction situation leading to the dopamine rush.

Addiction in the USA and global

Of course, you may think social media as an addiction seems as improbable as pointing fingers at violent video games or violence in features. But before social media, Americans suffered from roller discos, pet rocks, and bad hair choices. In the US, addiction is a pervasive issue affecting millions. The opioid crisis, particularly the fentanyl epidemic, has been devastating. Fentanyl, a synthetic opioid, is 50 to 100 times more potent than morphine. It's often mixed with other drugs, leading to an alarming increase in overdose death of 2023. Provisional data from the Centers for Disease Control and Prevention (CDC) indicate that an estimated 107,543 drug overdose deaths occurred in the United States, Illicit drug use is also a significant problem.[100] According to the latest 2021 data from the Substance Abuse and Mental Health Services Administration (SAMHSA), an estimated **46.3 million people aged 12 or older** in the U.S. had a substance use disorder (SUD) in the past year.[101] This includes the misuse of prescription drugs, which is another critical aspect of the addiction crisis. Mental health plays a crucial role in addiction. Many individuals with substance use disorders also suffer from co-occurring mental health conditions such as depression, anxiety, and post-traumatic stress disorder (PTSD). The interplay between mental health and addiction creates a complex challenge for healthcare providers.

Worldwide, addiction is a growing concern with varying substance preferences and impacts. In many parts of the world, alcohol remains a significant issue. As of 2023, alcohol-related deaths account for more than 2.6 million deaths annually worldwide, according to the World Health Organization (WHO). This represents around 4.7% of all global deaths, with a significant majority (approximately 2 million) occurring among men.[102] In Eastern Europe and Russia, alcohol addiction is particularly rampant, contributing to high mortality rates. Conversely, in regions like Southeast Asia and sub-Saharan Africa, the misuse of locally available substances, including cannabis and khat, poses significant health risks. The global rise in synthetic drug use, including methamphetamine and synthetic cannabinoids, is alarming. These substances are often more potent and dangerous, leading to severe health consequences and increased overdose deaths. Nowadays, addiction is digital.

The perils of dopamine dilemma

Excessive or compulsive use of devices and social networking is non-problematic for the majority of the population. Anything that sparks excitement in the brain could cause additive behaviors. If you're on one side of the fence looking at the problem, there are two people on the other side shouting and pointing fingers back, saying it's the continued ruin of Western civilization. Those little rewards your brain receives from recognition, notifications, mentions, and 'likes' kick your dopamine and add to the desire to give back as much as you received. Essentially, the brain rewires itself over time because of positive reinforcement.

That digital heroin by design is no accident. For every AI out there combatting the addiction process, bad actors use AI to create more algorithm prescriptions to hardwire your brain to keep you online and engaged. AI technology analyzes browsing habits, 'private' direct messages, and social media persuasions to enforce AI-curated contact, promoting harmful strategies. Remember those echo chambers? They keep happening because bad actors want to steal your intellectual likeness and absorb it into the collective. You can lose hours and days and withdraw from friends and family to maintain the new obsessions and trends. Big tech companies quarterly key performance indicators (KPIs) tell the truth about AI-manipulated content to kick-start your daily dopamine overdose.

The mastermind behind the scenes is AI

Here's where it gets even more pleasurable. AI algorithms, the masterminds behind the addictive nature of social media platforms, are capable of analyzing your behavior, preferences, and interactions to curate content most likely to keep you engaged. Every click, like, share, and response triggers more pleasurable brain candy. The AI prioritizes emotionally charged and sensational content, knowing this type of content generates more significant dopamine responses. For example, the powerful algorithms might notice that you spend more time on posts related to travel, and it will show you more travel content, making you feel special and catered to—bots will fuel artificial engagement—creating an even steadier stream of dopamine hits. It also tracks which posts get the most likes and comments, pushing similar content to the top of your feed. This continuous cycle of personalized, engaging content keeps you hooked, constantly seeking the next dopamine hit.

Another pandemic in the making

This isn't just an individual problem; it's becoming a societal issue. As more people, especially younger generations, spend increasing amounts of time on social media, the risk of widespread addiction grows. If you are unaware of recent events between January 2020 and May 2023, something worldwide happened that brought direct human interaction to an abrupt end while people forged new paths through the digital landscape. The internet highways became oceans to hold everyone's attention while the world rotated around the sun, and something unknown and unseen infected fewer and fewer people.

The economic burden became a staggering disaster, leading to a significant reduction in the Gross Domestic Product (GDP). Unemployment soared, and the longing for social interaction among humans became palpable. Digitalization surged by 6 percent, accelerating the need for more specialized AI technology. Forty-three percent of people in the US turned to enews, 42 percent to streaming services, 29 percent to online gaming, and social media use shot up by 50 percent. The list goes on, but a notable example of preoccupation during that year was the rise of other social media platforms. Onlyfans, for instance, saw a staggering 553 percent growth, with a $390 million increase in revenue.[103]

Dopamine also plays an important role in pain management. Pain aversion leads to quick fixes such as medications and dependency on addiction.

Pain averse society

The little thing that happened in 2020 isn't the only problem we're facing when it comes to distractions geared to keep your mind wandering so you don't face the pain of your daily existence. That sounds bleak; let's try something else. Even before 2020, over-the-counter prescription painkillers saw significant misuse. Streaming TV and movies have direct access to our ears and eyes. Big pharmaceutical companies target you through AI patterns in your searches and content watching to bring you endless commercials advertising medication for every weird ailment you feel throughout your long day. If you don't read the fine print or are busy looking at social media while the commercials pay in the background, you don't know the side effects are inherently worse than the symptoms.

In the US, 3 out of 10 women use sleep aids. Instead of asking, "Why aren't you sleeping?" Doctors will prescribe something to *make* you sleep, treating the symptoms instead of the problem. When you alleviate pain or submerge it along with anxiety, you ignore the warning signs. Physical or mental pain tries to tell you something important. Quieting the discomfort through prescriptions allows the underlying cause to linger, and you fail to identify the reason why you're suffering. Suppressing pain and anxiety cuts you off from experiencing emotion. Painkillers and sleep aids take away from the natural experiences; using medication for relief can lead to exacerbating the problem, putting your physical health at risk.

Interestingly, while your doctor happily fills those prescriptions you didn't need a few years ago, you can turn to online support and share your pain with others. Over 66 percent of adults reported taking prescribed medication in 2023. In 2020, people in the US filled 4.55 billion prescriptions. 2021-4.69 billion, in 2022-4.76 billion, and in 2023 people filled 4.83 billion prescriptions.[104] Are you tracking the numbers? Do you see what's happening with adults in the US?

The global pain epidemic

Researchers interviewed nearly 150,000 individuals across twenty-six countries and discovered a disturbing pattern: richer nations exhibit higher rates of anxiety than their poorer counterparts. This debilitating disorder is not only more prevalent but also more impairing in high-income countries. From 1990 to 2017, the number of new depression cases worldwide surged by 50 percent, with the most alarming increases seen in affluent regions, particularly North America.[105]

Physical pain is also on the rise, adding to this grim picture. Throughout my career, I have witnessed a growing number of patients suffering from widespread, inexplicable pain. Conditions such as complex regional pain syndrome, fibromyalgia, interstitial cystitis, myofascial pain syndrome, and pelvic pain syndrome are becoming increasingly common. When asked, "During the past four weeks, how often have you had bodily aches or pains?" a staggering 34 percent of Americans reported feeling pain "often" or "very often." This contrasts sharply with the much lower percentages in other countries: 19 percent in China, 18 percent in Japan, 13 percent in Switzerland, and only 11 percent in South Africa.[106]

Why are we so miserable?

Why, in a time of unprecedented wealth, freedom, technological progress, and medical advancement, are we unhappier and in more pain than ever? Social media and advertising often use AI to identify your problems and test the limits of your discomfort. This might not always cause pain, but it can lead to codependence. The AI isn't doing this on purpose; it simply follows the algorithms it was taught. By understanding these dynamics, you can learn to use technology to promote mental well-being.

Pain is a natural part of life. It signals when something is wrong and pushes us to take action. But in our quest to eliminate pain, you've become more dependent on external solutions like medications and distractions rather than addressing the root causes. Think about it: dopamine. Activities like drug use, social media engagement, and even eating can trigger pleasure in your brain. However, excessive stimulation can deplete dopamine levels, leading to a cycle of melancholy. Why? Cause your brain needs time to replenish the dopamine level. As a result, you feel sad and down even though you had 2 hours of TikTok or Reels watching. When you rely too heavily on quick fixes for pain—whether physical or emotional—you end up exacerbating the very problems you've trying to escape. The overuse of psychiatric drugs, stimulants, and sedatives is a testament to this vicious cycle. By understanding the root causes of your pain and taking steps to address them, you can break free from this cycle and improve your mental health.

Abundance creating a pain-averse generation

Social media reinforces the nature of pain or happiness, depending on which post you clicked, liked, or shared. Sex, food, and social interaction have push-pull elements that can weigh heavily on your emotions and mood. The idea of a potential future reward keeps the machines in use. The same goes for social media sites. One does not know how many likes a picture will get, who will like the picture, and when the picture will receive likes. The unknown outcome and the possibility of a desired outcome can keep users engaged with the sites.[107] The social circles are bigger than the town you live in; the entire planet can see what you're eating or photographing, saying, writing, and singing, you opened the

digital door, and the AI tsunami already knows everything about you, even if you're not sharing it. Low motivation is another significant issue. With everything available at the touch of a button, the drive to achieve and persevere diminishes. This generation, accustomed to instant gratification through digital means, often lacks the patience and resilience needed for long-term goals. A study from the National Institutes of Health (NIH) highlights that today's teens are less likely to engage in proactive problem-solving and exhibit lower motivation levels than previous generations.[108] Interestingly, in ancient times, people treated pain with pain!

Pain to treat pain

Throughout history, pain has been seen as a burden and a tool for growth and healing. In ancient Greece and other early civilizations, pain was sometimes used to treat pain, a practice rooted in the belief that enduring discomfort could lead to strength and recovery. Today, however, our society is largely pain-averse, seeking to eliminate discomfort through an array of medications and quick fixes. This avoidance may be doing more harm than good, fostering dependency on drugs with complex and often misunderstood long-term effects.

In ancient Greece, philosophers and physicians like Hippocrates viewed pain as a natural and necessary part of life. They believed that enduring pain could lead to resilience and healing. For example, treatments sometimes involve exposing patients to controlled pain to stimulate the body's natural healing processes. This perspective extended beyond physical pain to mental and emotional challenges, emphasizing the importance of facing and overcoming difficulties. Contrast this with our modern world, where pain is seen as something to be eradicated at all costs. Pain aversion has led to the widespread use of medications such as opioids, antidepressants, and anxiolytics. While these drugs can be effective in the short term, they often come with significant risks and side effects, particularly when used long-term.

Leaving pain to the analog world

Live the moment, support a healthy online community, and keep things in real life (IRL). When you see something that triggers your knee-jerk reaction to respond, take a moment before you comment. Write it, proofread it, and let the words sink in to see if it's worth a response. Is it true? Is it necessary? Is it kind? Don't compare yourself to others, you can't make anyone else happy from a million miles away without direct interaction. If you follow people online, make sure you're doing it because they share comment threads to positive thinking and represent lifestyles and attitudes that are real and not made up by trashy AI. If you see something stressful on social media, turn it off. Put down the phone *before* you climb into bed. Don't struggle alone; sometimes, you're not the only one who needs a break from social media.

The benefits of overcoming clickbait

You are not a smartphone. It might seem obvious or cliché, but you sometimes need reminding to avoid misinformation and clickbait. You need to ignore it. If you feel the need to keep looking, consider the source. Many bad actors employ AI to strengthen their hold on content. Your brain's filters come from your ability to recognize subtle hints to use logic over emotion. It's not harmless; it's by design. Spammy quizzes are rife with scams, AI-generated headlines, and fine-tuned, sleek, trashy text that probably has no human elements or validations.

Take time, vet your so-called friends and followers. Likely, more than you realize are bots and could be farming data instead of legitimate interaction. Keep your devices protected, change your passwords, and keep updated virus protections. You might get a text from a Nigerian Prince who needs help tracking Sasquatch. It might be legitimate, but no one puts Bigfoot in a box.

Laughing at foolishness: Are you SMART?

It would help to be cautious about the advice you give and receive, especially on social media. In today's world, social media is a breeding ground for unqualified individuals dispensing advice on subjects they know little about. Remember those Flat Earthers? The potential risks of following such advice are significant. Human language is incredibly nuanced, and words can have vastly different meanings depending on the context. For example, the phrase "I love you" can be said by a mother to her child, a brother to his sister, or a friend to another friend. Each instance carries a different weight and meaning based on the relationship and situation.

When someone on social media tells you, "You can do it; success is not a big deal," their context may differ entirely from yours. They're selling something, and it will cost you more. They will never know your name or care about your financial dilemmas. The importance of context in advice cannot be overstated. Someone who has found success in finance may need help understanding the challenges faced by someone working as a mechanic. Recognizing that the advice you receive must be relevant to your circumstances. If you don't, you might take risks based on misguided inspiration, and when things go wrong, you bear the consequences, not the person who gave the advice.

Consider this: people often post about how easy it is to become a business owner. But do you genuinely have the desire and the skills to run a business? These are decisive questions you need to ask yourself. You must follow such advice and understand your capabilities and goals to avoid deep trouble.

To navigate this maze of social media advice, you must apply "ME thinking," a blend of mindfulness and effortless thinking process. In the next chapter, you will learn this critical skill about ME (Mindfulness and Effortless) thinking. Look for posts presenting multiple perspectives and scenarios, not just one-sided. Life is complex and multifaceted, and your approach to it should be just as dynamic and unique. Yes, that does not mean you cannot give or take advice. But it would be best if you always gave it a grain of salt and a disclaimer that success is different for different people, and so is its Path. Remember, you don't want to end up like a Flat-Earther—laughed at for clearly misguided beliefs. Just as you would unsympathetically evaluate any significant decision in life, apply the same scrutiny to the advice you encounter online. This way, you can navigate the digital landscape wisely and make decisions that truly benefit you.

Adopting a thoughtful and discerning approach to social

media advice means you're not just following the crowd. Instead, you're making informed, intelligent choices that propel you forward. This empowerment is the key to navigating the digital landscape wisely and making decisions that truly benefit you.

Your Guide to a Digital Detox

It's time to use the AI tools you can access without spending more time and money on things you probably don't need. Currently, your smartphone has features to help eliminate the toxic embedded nature perpetuating the social media scene. If you're conscious of your need to stay connected, even if someone hid your phone earlier and you had to revert to your animal side to coax that person to give it back, you know it's important to detox and turn it off.

Tailor Your Detox: A Personal Touch

Don't settle for a generic approach. Design a detox plan that's tailored to your unique needs. Which apps are your downfalls? Is it the endless vortex of TikTok, the meticulously crafted perfection of Instagram, or the dopamine surge of X's or Threads notifications? Identify your main offenders and start chipping away at your time on these platforms. It's like slowly but surely breaking free from a sugary indulgence.

Your Phone: A Spy in Your Pocket

Unless you actively turned off your voice-assistant features on your smartphone, it's always listening to you and your conversation. It's constantly monitoring your life, aware of your struggles with self-image and low self-esteem, and your sleep problems. It knows when you're awake and when you've been good or naughty. It doesn't have any feelings one way or the other, but it won't stop unless you can teach it to ignore you or turn off the smartphone altogether.

Baby Steps to Big Changes: Gradual Reduction

Don't stop immediately-your brain might revolt! Instead, ease into it by setting daily or weekly limits for each app. Disable or uninstall superfluous apps. Your app setting can tell you how long apps are idle. Start with uninstalling those first. It's like training for a marathon; you can't run 26 miles on your first day. Start with a 10-minute daily reduction and gradually increase it as you strengthen. This gives your brain time to adjust to lower dopamine levels, making the detox process less painful and more sustainable.

Offline Adventures Await: Mindful Substitution

Social media is not everything; ask anyone over 40, and they'll tell you about the old days when dial-up internet cost 12.95 USD a day and you had access to a whopping 56 Kbps super-speed. Downloading a 3-minute file took a few hours, and streaming videos or music was impossible. They can tell you what it's like to have an entire world waiting outside, offline. Don't feel obligated to take pictures of every meal you eat at a restaurant. Enjoy a tranquil walk through the park or nature trail without the pressure of streaming it live. Until you catch great shots of Sasquatch hitchhiking or aliens asking for directions, the views you get on an average day will be fewer than 10. But does that really matter?

Give Your Brain a Break: Scheduled Breaks

Do you have tinnitus and don't know it because you're surrounded by digital noise and can't hear clearly? Take a break, disconnect, and reevaluate your decision to buy roller skates for your dog. You need to actively shut it out because it wants to keep you connected. Taking breaks from social media will help your brain reboot and decompress. It's a lot like exercise; it hurts at first, but after a while, it feels better, and your body and brain seek different ways to trigger dopamine, like spending time with loved ones, engaging in hobbies, or simply enjoying a good book. Does it seem cliché to feel good about possibly keeping secrets from your online followers?

Silence the Noise: Notification Management

Turn off the notifications. You don't need reminders when your online game says it's time to harvest the digital fruit to make the digital wine to sell at the digital market. You are not a machine, but the AI in your smartphone thinks you are irresponsible, needs reminders to stay connected, and actively sends you pings and buzzes from every app on your platform. It's exhausting and designed to keep you awake and distracted. Take time to stop meaningless notifications. It won't take all day, only as long as the amount of apps you have on your phone.

Unplug Before Bedtime: Evening Routine

Your subconscious mind needs to clear your daily brain browser history, but it can't do its job if you keep active through your sleepy time because that phone is shining blue light in your eyes. Plenty of apps are designed to help your brain reset; if you can't sleep and think you need prescriptions, use a day you don't have to wake up early to experiment with different online sleep features if you must stay connected; why not let AI help your subconscious find a channel that turned off the conscious brain so it can do the necessary maintenance?

The Marathon, Not the Sprint: Long-Term Sustainability

Remember, this isn't a race. Focus on creating lasting habits, not quick fixes. Celebrate your small victories along the way and don't beat yourself up if you slip up occasionally. It's all about progress, not perfection.

Get Expert Help: Professional Guidance

If you're struggling to break free from the digital chains, don't hesitate to seek professional help. Therapists specializing in digital addiction can provide personalized strategies and support to help you regain control of your life.

The Chinese Initiative towards Digital Detox

To combat gaming addiction and promote healthier lifestyles among its youth, China has implemented strict regulations limiting online video game play to just two hours per week for children. This policy, introduced by the Chinese government, aims to reduce the negative impacts of excessive gaming on young minds, such as poor academic performance and physical health issues.[109] These restrictions are part of a broader initiative to encourage a digital media detox, helping children to balance their screen time with other activities that foster physical and mental well-being. By promoting moderated use of digital technology, China hopes to set a precedent for healthier, more balanced lifestyles among its younger population, addressing concerns over digital addiction and its long-term consequences. This approach serves as a reminder of the importance of digital media detox practices globally, emphasizing the need for structured screen time to maintain a healthy balance in your increasingly digital world. It also highlights the global relevance of the issue, as many countries are grappling with similar concerns and could benefit from China's example.

The unexamined life is not worth living

Social media addiction is now officially recognized as a diagnosable disorder, and it all ties back to dopamine, your brain's little cheerleader that keeps you motivated and happy.[110] When you eat your favorite food or win a game, dopamine gives you that feel-good rush. Social media platforms have cleverly tapped into this by rewarding you with likes, comments, and shares. Every time you get one, your brain releases a tiny bit of dopamine, making you feel great and eager for more. The reward loop keeps you coming back.

But here's the kicker: social media isn't just about dopamine hits. AI algorithms designed to keep you hooked don't have feelings, so switching off won't matter to them either way. These algorithms prioritize emotionally charged content because it generates more significant dopamine responses by design. The more you engage, the more your brain craves these dopamine boosts, leading to compulsive behavior and making everyday activities seem less enjoyable. Stay aware, set boundaries, and remember that the control is in your hands. You can manage your digital detox to reset dopamine reserve in your brain and find joy in real-world interactions to become the best version of the humAIn.

While the media and some companies might spread fear and misinformation to keep these tools exclusive and profit-driven, the next chapter will guide you on how to navigate these challenges. Discover how to overcome fear and misinformation, empowering yourself by utilizing ME thinking and becoming the next generation of innovative humAIn.

Challenge your thinking

- Can you implement the digital detox guidelines mentioned in this chapter? It takes work to do something suddenly. You will need help and figure out how your family and friends can help. One way to deal with this is to avoid taking the smartphone to your room. You can rule that all family members should keep their phones in the living room 2 days a week and gradually increase. Parents should be leading by example. Read a book, and if you don't want to 'read' the book, and then use AI to find audiobooks you can download.

chapter 9

BECOMING THE WISE HUMAIN

Do not believe in anything simply because you have heard it. Do not believe in anything simply because it is spoken and rumored by many. Do not believe in anything simply because it is found written in your religious books. Do not believe in anything merely on the authority of your teachers and elders. Do not believe in traditions because they have been handed down for many generations. But after observation and analysis, when you find that anything agrees with reason and is conducive to the good and benefit of one and all, then accept it and live up to it. - Siddhartha Gautama (Buddha)

Go with the FLOW

Once upon a time, in a peaceful village nestled within a great kingdom, there lived a humble farmer and his beloved son. The farmer's wife had tragically passed away during childbirth, leaving him to raise his son alone. Despite his lack of formal education, the farmer was a well-read man, immersing himself in books about philosophy and history. His wisdom made him a cherished figure among the villagers, who often sought his counsel in times of need.

The farmer and his son were poor, relying on a single horse to plow their land and sustain their livelihood. One day, the son took their horse out to graze, but the horse bolted into the wilderness. Devastated, neighbors and friends rushed to console the farmer, lamenting his misfortune and fearing for his future. But the farmer, serene and untroubled, simply said, "Whatever happens, happens for a great reason."

The next day, the runaway horse returned, bringing with it a wild stallion. The neighbors were ecstatic, congratulating the farmer on his newfound luck. "Now you have two horses! Your fortunes will surely improve," they exclaimed. Yet again, the farmer replied calmly, "Whatever happens, happens for a great reason."

As time passed, the farmer's son attempted to ride the wild horse but was thrown off, breaking his leg. The villagers gathered; their faces etched with sorrow. "This is terrible," they sympathized. "How will you manage now?" Unfazed, the farmer reassured them, "Whatever happens, happens for a great reason."

Soon after, a war broke out, and the kingdom's soldiers came to conscript able-bodied young men. They took the sons of all the villagers but left the farmer's son due to his injury. The neighbors, now fearful for their own children, returned to the farmer, marveling at his fortune. "You were right," they admitted. "Your son is safe because of his broken leg. What should we do? How should we think?" The farmer smiled gently and said, "Whatever happens, happens for a great reason.

The moral of the story is profound: We often react to life's events without understanding their true significance. Every occurrence is neutral until we assign meaning to it, labeling it as good or bad. This tale teaches us to embrace life's events with calm acceptance, knowing that there may be a greater reason behind each twist and turn. The farmer in this story has mastered the power of mindfulness and effortless thinking. I invite you to discover this power in this chapter.

If GOD wants to punish you, your wish gets granted

As you have seen in Chapter 6 of *Polarization*, we often categorize things into binary buckets: good and bad, right and wrong, fair and unfair. This habit stems from our brain's need to conserve energy and make quick decisions. While this method of thinking has served us well historically, it is becoming less effective in today's world, especially as we become more reliant on AI. If you continue to categorize information simplistically, you will miss out on the nuanced understanding necessary to navigate the complexities of modern life. It will just be noise and nothing else.

In our lives, we react to events. However, these events have no inherent meaning until we assign one to them. Controlling our reactions is crucial because the meanings we assign can often lead us astray. Think of it this way: in your life, there are only two true categories: being alive and dead. Everything else is a matter of perception. I have seen people with millions of dollars in their bank accounts who live miserable lives, constantly wanting more and feeling depressed and suicidal. Their mental health is in shambles, and they lack meaningful connections. Conversely, I have seen people with nothing—no savings, no material wealth—yet they lead rich lives filled with family, friends, and respect. Their children care for them, and they find joy in their daily interactions.

Life is not just about pursuing things but about understanding how your brain works and managing the information around you. The purpose of AI is to provide you with the information at your fingertips. As you seek happiness and the ultimate utopia, you must realize that ultimate happiness doesn't exist as a permanent state. Life is a constant flow of experiences, and your brain continually searches for ways to categorize these experiences. You need to learn how to manage this process to live a joyful and calm life, like the farmer who took things as they came without trying to control or change them.

In this chapter, you will learn how to break free from the cycle of pursuit of happiness and manage your reactions to life's events in the age of AI. If you don't learn this, you risk falling into the patterns of declining mental health that we have discussed in chapter seven. Mental health issues are not just a national concern; they are a global crisis exacerbated by the pervasive influence of social media. The strategies you will learn in this chapter are designed to help you live a fulfilling life with AI. By understanding your brain's tendencies and learning to manage the information it processes, you can achieve a joyful, balanced and serene state of mind. It all starts not just with an open mind, but with active open thinking.

Active Open Thinking

Are you an open-minded person? The most closed-minded person I know will answer YES to the question if I ask are you open minded. Active open thinking is a concept that emphasizes an intentional and deliberate approach to processing information.[111] It involves a mindset where you remain receptive to new ideas. Your assumptions, and actively seek diverse perspectives. This way of thinking contrasts with the brain's natural tendency to make quick, automatic judgments to conserve energy. Instead of defaulting to habitual thought patterns, you consciously engage in a thoughtful and reflective process, ensuring that your conclusions are well-informed and balanced.

To truly understand active open thinking, picture a time when you encountered a problem at work that seemed straightforward. Perhaps you instinctively reached for a familiar solution without much thought. However, practicing active open thinking would mean pausing to consider alternative solutions, questioning your initial assumptions, and seeking input from colleagues with different expertise. This approach might reveal a more innovative or effective solution you would have missed if you had relied solely on automatic, habitual thinking.

One real-life example of active open thinking comes from the medical field. Consider a doctor diagnosing a patient with a common set of symptoms. The doctor could quickly default to a routine diagnosis based on past cases. However, by engaging in active open thinking, the doctor would question this initial diagnosis, consider fewer common conditions, and consult with other specialists. This thorough process could lead to a more accurate diagnosis, ultimately improving patient outcomes.

Another example can be found in the realm of personal relationships. Suppose you have a disagreement with a close friend. Your brain might automatically categorize the situation based on past conflicts, leading you to react defensively. Active open thinking, however, encourages you to step back, consider the other person's perspective, and explore the underlying reasons for the disagreement. This mindful approach can foster better understanding and strengthen your relationship.

Active open thinking is not just about being open-minded; it's about being proactive in your pursuit of knowledge and understanding. It involves a deliberate effort to seek out new information, especially from sources that challenge your existing beliefs. This doesn't mean you have to agree with every new perspective, but it does mean considering it fairly before forming a judgment. For instance, in the world of business, leaders who practice active open thinking are more likely to innovate and adapt to changing markets. Take the example of Netflix, which transitioned from a DVD rental service to a streaming giant.

The company's leaders didn't just stick to their original business model; they continuously questioned their assumptions, explored new technologies, and adapted their strategies based on emerging trends and consumer behaviors. This willingness to think actively and openly allowed Netflix to stay ahead of the competition and redefine the entertainment industry.

Your brain's default mode is to take shortcuts, using heuristics and biases to quickly make sense of the world. This energy-saving mechanism works well in routine situations but can lead to errors in complex or novel scenarios. Active open thinking is about being aware of these cognitive shortcuts and intentionally counteracting them. It requires effort, but the rewards are significant: better decisions, more creative solutions, and deeper understanding.

One way to cultivate active open thinking is through mindfulness practices. Mindfulness helps you become more aware of your thought processes, enabling you to notice when you're falling into automatic patterns. By practicing mindfulness, you can develop the habit of pausing before making decisions, allowing yourself the space to consider alternative viewpoints and gather more information.

Active open thinking also involves a commitment to lifelong learning. In a rapidly changing world, staying informed and adaptable is crucial. This means continuously seeking out new knowledge through formal education, reading, or engaging with others with different experiences and expertise. The definition of learning is a change in behavior. Ask yourself what you truly changed in your actions and behavior last time? When did you figure out that your thinking was totally wrong and changed your actions? If those answers are not frequent, then you are probably not practicing much of active open thinking. Consider the story of Steve Jobs, who attributed much of his creative success to his curiosity and willingness to learn from various fields. Jobs took calligraphy classes, studied design, and constantly explored new technologies. This active pursuit of diverse knowledge enabled him to create groundbreaking products that revolutionized multiple industries.

In today's age of AI, active open thinking is crucial, especially as AI algorithms curate the information we see on social media, subtly shaping our perceptions and interactions with the world. AI-driven platforms prioritize content that keeps us engaged, often amplifying sensational or emotionally charged posts, which can blur the line between real and fake news. For instance, you might notice your social media feed filled with alarming headlines or polarizing opinions designed to capture your attention. Without active open thinking, your brain's natural tendency to take cognitive shortcuts can lead you to accept misinformation. By actively questioning, seeking diverse perspectives, and understanding how your mind processes information, you can better navigate the sea of misinformation. This awareness helps you discern truth from falsehood and fosters more meaningful and authentic connections in a world increasingly mediated by technology. Therefore, ME thinking becomes the most important lesson that your education system never taught you.

ME Thinking

In our daily lives, our brains employ two distinct processes for thinking: M (Mindfulness) and E (Effortless).[112] These processes illustrate how we navigate the complexities of our environment. Let's delve into these two modes of thinking, using practical examples to make them relevant.

Think about the first time you learned to drive a car. Initially, every action required your full attention. You had to think about pressing the brake, turning the steering wheel, checking your mirrors, and coordinating your movements. This deliberate, effortful thinking is what we call M thinking. It's when you are fully engaged, conscious, and using active judgment to process information. You are mindful of every detail and make decisions based on careful consideration.

Now, contrast that with how you drive today after years of experience. You might get into your car, start the engine, and drive to work while chatting with a friend or listening to music. You arrive at your destination and realize you don't remember every turn or stop you made along the way. This is E (Effortless) thinking in action. Your brain has automated the driving process to the point where it requires little conscious effort. The skills and judgments you developed through mindful practice have become second nature, allowing you to perform tasks without actively thinking about them.

This transition from M to E thinking is a hallmark of how our brains manage tasks efficiently. When we first encounter a new situation or learn a new skill, we rely on M thinking. We analyze, deliberate, and make conscious decisions. Over time, with repetition and practice, these tasks become automatic and shift to E thinking, freeing up cognitive resources for other activities.

However, the ease and efficiency of E thinking can be a double-edged sword, especially in today's digital age of AI. Social media platforms and the constant flow of information can overwhelm our cognitive systems. When we scroll through our feeds, our brains often operate in E mode, quickly processing headlines, images, and snippets of text without much scrutiny. This can lead to the rapid acceptance of misinformation and biased content because our brains are not engaging in the mindful analysis that M thinking provides.

Let's take a real-world scenario: you're scrolling through your social media feed and come across a sensational news headline. In E mode, you might accept the headline at face value and share it with friends without verifying its accuracy. However, if you engage in M thinking, you will pause, consider the source, cross-check the information with other credible sources, and then form a judgment. This mindful approach helps you filter out misinformation and make more informed decisions.

The consequences of relying too heavily on E thinking in the context of social media and news consumption can be significant. Misinformation can spread rapidly, influencing public opinion and behavior in harmful ways. For example, during the COVID-19 pandemic, false information about treatments and vaccines circulated widely on social media, leading some people to make dangerous health decisions based on misinformation.

Understanding the balance between M and E thinking is crucial for navigating the modern world. By recognizing when you are in E mode, you can consciously shift to M mode when the situation demands it. This means being more critical of the information you consume, questioning its validity, and seeking multiple perspectives before forming an opinion. Consider the example of a professional setting. If you are working on a complex project that requires critical thinking and problem-solving. If you approach it in E mode, you might miss important details or make hasty decisions based on incomplete information. However, by deliberately engaging in M thinking, you can analyze the problem from multiple angles, consider various solutions, and make well-informed decisions that lead to better outcomes. In your personal life, practicing M thinking can improve your relationships and overall well-being. For instance, during conversations with loved ones, being fully present and actively listening (M thinking) rather than multitasking and giving partial attention (E thinking) can strengthen your connections and enhance mutual understanding.

Recognizing the difference between M (Mindfulness) and E (Effortless) thinking can help you better manage the flood of information in today's digital age. By training yourself to switch to M thinking when needed, you can critically evaluate information, avoid the pitfalls of misinformation, and lead a more mindful, intentional life. Remember, while E thinking helps us navigate routine tasks effortlessly, M thinking equips us to tackle complex problems and make informed decisions that shape our lives positively. In the following sections, we will learn details on how you can train your brain and your AI in the M and E thinking process.

ME Thinking Process: Mindful and Effortless

In this section, we're diving into the concept of M thinking— Mindfulness thinking—and why it's essential in our modern, AI-driven world.

M for Mindful Thinking in ME thinking

M thinking is all about being present, deliberate, and reflective. It's about engaging actively with the information at hand, analyzing it critically, and making thoughtful decisions. In today's age of AI and digital information overload, M thinking is more crucial than ever. Let's dive into the checklist for ME thinking using M thinking and explore how you can apply it in your daily life.

Checklist for ME: For M

I created an easy way for you to navigate through the M part of ME thinking. It is called MINDSET

- M: Metacognition

- I: Information source motivation

- N: Navigate Opposite information

- D: Determine the worst that can happen

- S: Statistical Thinking

- E: Evaluate Higher Purpose

- T: Traceback from End Point

Metacognition (Thinking about Thinking)

Think of metacognition as a mental mirror. It's about being aware of your thought processes and evaluating how you're thinking. For example, when you're reading an article online, ask yourself:

"Why am I accepting this information as true? What biases might I have?" By questioning your thought process, you become more mindful and less likely to fall for misinformation. You question your thinking and thinking about your thinking make you think more.

Information Source Motivation

Understanding the motivation behind an information source is critical for accurately assessing its reliability. When you come across new information, the first step is to consider the source. Who is providing this information? Is it a credible and reputable source, or is it potentially biased? For example, if a well-known medical journal publishes a study, it is likely more reliable than a random blog post. Next, evaluate the applicability of the information. Ask yourself if this information can be applied universally or is specific to particular contexts. For instance, a dietary recommendation might work well for a particular group of people but may not be suitable for everyone. Lastly, consider the evidence and reproducibility of the information. Is the information based on solid, scientific evidence? Can the results be reproduced by other researchers or studies? Using the example of a new diet trend, it's essential to check if the diet plan is backed by scientific research and peer-reviewed studies or if it's merely a marketing gimmick aimed at selling products. By asking these questions, you can better navigate the sea of information and make more informed decisions.

Navigate Opposite Information

Think you're a detective, piecing together a story from different sources. Navigating opposite information involves seeking out opposing viewpoints and comparing them to find the most accurate picture. Don't just rely on one source if you're researching a health trend. Look for studies, expert opinions, and real-life testimonials to get a comprehensive understanding. This is exactly what scientists do when they create a research hypothesis, which is trying to find evidence in the opposite information so they can reject it. All scientists are trained to create a null hypothesis which is precisely to assume that opposite information is correct. Then, we seek evidence that can support the opposite information and if that fails, we reject the null hypothesis and accept the fact. You may be thinking that you are not a scientist. But the current time is so dire that you must think like one to survive in the age of misinformation and disinformation. You learned in detail about triangulating or navigating opposite information in chapter 6 to detect misinformation.

Determine the Worst That Can Happen

Fear of the unknown can be paralyzing. Ask yourself, "What's the worst that can happen?" Often, the potential adverse outcomes are less severe than we imagine. This perspective can give you the courage to take calculated risks. If you're considering a career change, evaluate the worst-case scenario and plan accordingly. The worst that can happen might be temporary discomfort, which is often outweighed by potential growth and satisfaction. Even though your worst may be significant for you at that very moment; with time fear lessens and you learn to adapt. Our brain is highly adaptive, and our resilience has been built over millions of years. So, you can overcome any worst situation as life always moves on.

Statistical Thinking

Numbers don't lie, but they can be misleading if not interpreted correctly. Statistical thinking involves understanding probabilities and risks. For example, if you're reading about a new medication, look at the statistics: What is the success rate? What are the potential side effects? What is the number needed to treat? By understanding the numbers, you can make more informed decisions. AI can help decipher large amounts of data and statistics. Use AI to guide you in finding the relevant statistics for your situational needs.

Evaluate Higher Purpose Beyond Yourself

Connecting your actions and decisions to a higher purpose can provide a sense of clarity and motivation that drives you through even the most challenging tasks. Anchoring your efforts to a broader goal, such as contributing to your community, protecting the environment, or advancing human knowledge, can help you stay focused and energized. Let's dive into how this works and why it's so powerful. When you're working on a challenging project at work, it's easy to get bogged down by the details and lose sight of the bigger picture. But if you remind yourself of the more significant impact your work will have, it can reignite your motivation. For instance, if you're developing a new product, think about how it will improve people's lives, create jobs, or drive innovation in your industry. This connection to a higher purpose transforms mundane tasks into meaningful actions.

Take another example: if you're involved in community service, it's not just about the hours you put in. It's about the lives you touch and the positive changes you bring to your community. Whether you're organizing a local clean-up to protect the environment or mentoring young students to help them achieve their dreams, keeping the higher purpose in mind gives you the perseverance to push through challenges. Even in your personal life, anchoring your actions to a higher purpose can be transformative. If you're trying to lead a healthier lifestyle, consider the broader implications: living longer to spend more time with loved ones, setting a positive example for your children, or reducing your environmental footprint. These higher purposes make it easier to stick to your goals, even when temptation or discouragement strikes.

Ultimately, by aligning your daily actions with a higher purpose, you create a powerful source of intrinsic motivation. This approach not only helps you achieve your goals but also brings a more profound sense of fulfillment and satisfaction. So, the next time you face a challenging task, take a moment to reflect on the bigger picture and let that higher purpose guide you forward.

Trackback from your endpoint

To navigate life's challenges and achieve your goals, it's crucial to start with the endgoal in mind. Picture what you want to achieve and then work backward to identify the necessary steps to get there. This method helps you create a clear and actionable plan. Let's dive into how you can apply this strategy effectively.

First, ask yourself, "What do I need to achieve?" Defining your ultimate goal is essential. Whether it's running a marathon, launching a business, or mastering a new skill, having a clear objective keeps you focused and motivated. For instance, if your goal is to run a marathon, the end goal isn't just the race day but also completing it at a specific time or simply crossing the finish line.

Next, consider, "What do I need to learn?" The next step is understanding the knowledge and skills required to achieve your goal. In the marathon example, you would need to learn about proper running techniques, nutrition, injury prevention, and the mental strategies that long-distance runners use. This might involve reading books, watching instructional videos, or consulting with experienced runners and coaches.

Finally, ponder, "How do I learn it?" Identifying the best way to acquire the necessary knowledge and skills is crucial. This might include enrolling in a training program, joining a running club, or following a structured training plan. You could also seek mentorship from experienced runners who can provide guidance and support. For the marathon example, you might break down your learning process into smaller, manageable chunks: improving your endurance, learning about proper nutrition, and developing a detailed training schedule.

This approach can be applied to any goal. If you're launching a business, start by identifying your end goal, such as having a profitable enterprise in a specific industry. Then, determine what you need to learn: market research, business planning, financial management, and marketing strategies. Finally, figure out how to learn it: taking business courses, attending workshops, networking with other entrepreneurs, and seeking mentorship.

By starting with the end in mind and working backward, you create a roadmap that guides your journey. This method helps you stay organized, focused, and motivated, ensuring that you take the necessary steps to achieve your goals. It also allows you to anticipate and prepare for potential challenges in advance.

Now, we will look at the E (effortless) thinking which has fewer steps as it is the autopilot for our cognition.

E for Effortless Thinking in ME thinking
Effortless Thinking in the Age of AI

Our brains are constantly bombarded with information in today's fast-paced digital world. The barrage of misinformation can trigger the release of dopamine, a neurotransmitter that plays a crucial role in our brain's reward system. The allure of dopamine can make us crave the instant gratification provided by TikTok or Instagram Reels, where every like, share, and comment gives us a little rush of happiness. However, while this dopamine-driven "effortless thinking" can make life enjoyable, it's essential to balance it with mindful thinking to filter the vast amounts of information we receive.

Effortless thinking is necessary because it's simply impractical to scrutinize every piece of information that comes our way. We need to rely on quick, automatic responses to navigate our daily lives efficiently. For instance, when scrolling through social media and seeing a couple enjoying a romantic dinner, it's natural to think, "How lovely!" But with mindful thinking, you might realize they are performing for the camera, adding a layer of pressure and possibly detracting from their genuine enjoyment. By training your mind to balance these thought processes, your effortless thinking can incorporate elements of mindfulness, making you more discerning without overburdening your cognitive resources to achieve FLOW.

Psychologist Mihaly Csikszentmihalyi introduced the concept of "FLOW," a state where people are so engrossed in an activity that they lose track of time.[66] Achieving flow requires a blend of effortless thinking and emotional engagement. When you're in flow, you're not just completing a task; you're immersed in it, experiencing a seamless connection between mind and body. Think of an artist painting a masterpiece, a musician composing a symphony, or an athlete at the peak of their game. They aren't merely going through the motions; they are fully absorbed in the moment, driven by passion and expertise. To achieve flow, you need to engage your emotions and let your natural abilities shine rather than overthinking every move.

Flow can be achieved in various aspects of life, from professional work to personal hobbies. Consider a writer who becomes so involved in their story that hours pass without them noticing. Their fingers fly over the keyboard as if on autopilot, guided by the narrative in their mind. Or think about a software developer writing code for an innovative project. They might start the day with a clear objective, but as they delve into the problem-solving process, they become completely absorbed, losing track of time and feeling a deep sense of fulfillment. This state of flow is where effortless thinking meets emotional investment, creating an optimal experience that is both productive and deeply satisfying.

To cultivate this balance between effortless and mindfulness thinking, especially in the age of AI, it's crucial to train your brain. This involves building emotional intelligence and learning how to process data effectively. In a world where AI algorithms and social media platforms can easily manipulate information, a disciplined approach to consuming and interpreting content is vital. Start by curating your digital environment. Regularly clean up your social media feeds to ensure you follow people and pages that provide value and positivity. Eliminate connections that propagate negativity or one-sided viewpoints, as these can distort your perception and contribute to misinformation.

Creating a supportive and diverse social network is another essential step. Surround yourself with individuals who challenge your thinking and encourage open-mindedness. These are people who value doubt and are open to different perspectives, fostering an environment of kindness rather than polarization. By doing so, you build a tribe supporting your mental and emotional well-being, helping you navigate the digital landscape more clearly.

Effortless thinking is essential for enjoying life and making quick, everyday decisions without becoming overwhelmed. However, it must be balanced with mindful thinking to navigate the complexities of the modern information age effectively. By integrating mindfulness into your effortless thinking processes, you can enjoy the flow of life while remaining vigilant against misinformation and manipulation. This balanced approach enhances your mental resilience and empowers you to live a more fulfilling and informed life.

Effortless thinking allows you to enjoy the moment and make quick decisions, but it's crucial to complement it with mindful thinking to ensure you process information accurately. By training your brain, curating your digital environment, building a supportive social network, and personalizing your AI algorithms, you can create a balanced approach to thinking that enhances your happiness and understanding of the world. Embrace this balance, and you'll find yourself better equipped to navigate the digital age with confidence and clarity.

Steps to Master E-Thinking

Here's an acronym to help you remember the steps for transitioning from mindful to effortless thinking when dealing with AI fears: CALM.

CALM

- C: Collect Evidence
- A: Activate Effortless Mode
- L: Lodge Positive Beliefs
- M: Maintain a Balanced Digital Diet

Let us assume that you are worried that AI will take away your job. Now, using CALM technique, you can learn to deal with the fear and anxiety of losing your job.

Step C. Collect Evidence

After engaging in mindfulness (M) thinking and finding solid evidence that AI won't take your job, internalize this information. Understand that while AI might change how some tasks are performed, it won't replace the essential human aspects of your job. For example, if you're a nurse, focus on the compassionate care and complex decision-making that AI cannot replicate. Look at the current job placement, openings and statistics to collect enough evidence.

Step A. Activate Effortless Mode

Allow your brain to shift to a more relaxed, automatic mode when you encounter similar information in the future. You've done the hard work of researching and analyzing the evidence, so trust your findings. This means that when you see another alarming headline about AI, you can effortlessly recall your earlier research and stay calm.

Step L. Lodge Positive Beliefs

Each time you see a fear-inducing headline about AI and jobs, remind yourself of the evidence you've gathered. This reinforcement helps maintain a balanced perspective. If a news article claims that AI will dominate your field, remember the studies and expert opinions that emphasize the irreplaceable human elements of your work.

Step M. Maintain a Balanced Digital Diet

Regularly review your social media and news feeds. Ensure that the information you consume is balanced, not just sensationalist content designed to provoke fear. This practice helps keep your effortless thinking mode healthy and productive. For instance, follow credible sources that provide balanced insights about AI rather than those that sensationalize its impact. Make a habit of cleaning unwanted friends and followers once a month. Getting false kudos from people who laugh on your back is not worth your time.

I provided an appendix at the end for you to practice the MINDSET and CALM techniques. I have counseled thousands of students and professionals to achieve their goals with these techniques and you can do it too!

By following the MINDSET and CALM steps, you can effectively manage any fear. If you encounter any piece of information, ask yourself, "Am I in M or E mode? Should I change my thinking to M mode?" Maybe you will learn to detect specific information that you need to switch to M mode. You'll learn to balance mindful analysis with effortless information processing, allowing you to stay informed and calm in the face of rapid technological changes. This balanced approach will help you navigate the AI revolution and empower you for growth and improvement in your professional life. Employing this ME thinking you can be the best version of humAIn.

In the next chapter, we'll explore how AI seamlessly integrates into our daily lives, transforming us into more efficient, productive, and incredible beings. This exciting Aivolution promises to enhance every aspect of our existence, making tasks easier and enriching our experiences. You'll discover practical ways to harness the power of AI while staying in control, ensuring you remain the best version of humAIn.

Challenge your Thinking

- Do you find yourself on autopilot, mindlessly scrolling through news and social media? Try the Mindfulness of ME thinking and see how quickly you start to feel exhausted.

- Effortless thinking can be beneficial if you first use mindfulness to focus on the type of information you're consuming. In an age dominated by AI algorithms, resisting the dopamine hooks that drain your brain's capacity to enjoy content is challenging.

chapter 10

HUMAIN IN AIVOLUTION

Chance favors the prepared mind. - Louis Pastor

Let's look into these two interesting stories:

The invention of fire

Once upon a time, in the ancient forests, two tribes lived close to each other. The first tribe, known as the Ember Clan, made a discovery that would change their lives forever. One stormy night, lightning struck a tree, setting it on fire. The tribe, led by their wise elder Mira, watched the flames with awe and curiosity. Always eager to learn from nature, Mira saw potential in this strange new force.

Under Mira's guidance, the Ember Clan learned to control the fire. They used it to cook food, keep warm, and protect themselves from predators. This new knowledge brought many benefits. Cooked food reduced diseases from raw meat, the warmth from the fire helped them survive cold winters, and the light extended their day, allowing for more productive activities. Around the campfire, they shared stories, passed down knowledge, and strengthened their bonds. The Ember Clan began to thrive.

Meanwhile, the Frost Clan, their neighbors, watched with a mix of fear and skepticism. Their leader, Koran, saw the fire as dangerous and warned his people to stay away from it. The Frost Clan continued their traditional ways, using animal skins for warmth and eating raw food. They viewed the Ember Clan's new practices as reckless.

As winters grew harsher, the Frost Clan struggled. They lost members to the cold and diseases that could have been prevented. Without fire to guide them during long nights, gathering food became harder. Desperation set in. Some members of the Frost Clan, driven by harsh conditions, sought the warmth and safety of the Ember Clan's fire. Those who adapted and embraced the change found new opportunities for survival and prosperity. Those who held onto the old ways gradually disappeared, highlighting the importance of adaptation.

The steam age ascendant

In the heart of the bustling industrial city of Ironbridge, during the transformative Steam Age, two families, the Lockes and the Ashworths, lived close to each other. Elias Locke, the forward-thinking head of the Locke family, saw immense potential in the new steam power and mechanized production technologies. He boldly invested in these innovations, turning his small workshop into a thriving factory. Elias's commitment to change and progress brought his family great prosperity and influence.

The Lockes, with their determined spirit, mastered the new machines and hired skilled workers to increase their production. Their focus on efficiency and innovation allowed them to export their products far and wide, making them pioneers of the Steam Age. Their success showed the limitless opportunities that came with embracing technological advancement.

In contrast, the Ashworths, led by the cautious patriarch Arthur, viewed the rise of machines with fear. Arthur strongly believed in the traditional methods of hand craftsmanship passed down through generations. He worried that machines would replace skilled artisans and lower the quality of their carefully made products. Despite the changes around them, the Ashworths held on to their traditions.

As Ironbridge changed dramatically with new factories, railroads, and businesses, the Locke family continued to grow. They expanded their operations, creating many jobs and opportunities for the townspeople. Meanwhile, the Ashworths' workshop struggled. Their handcrafted items couldn't compete with the high volume and low cost of machine-made goods. Orders declined, and their loyal customers turned to more efficient suppliers. Arthur Ashworth's refusal to adapt led his family to the brink of financial ruin, while the Lockes thrived, showing the power of embracing innovation.

As we stand on the brink of the AIvolution or AI revolution, these stories serve as powerful reminders of the importance of adaptation. Just as the Ember Clan thrived with fire and the

Lockes family soared with industrialization, , those who embrace AI will find new opportunities and growth. The lessons from these stories are clear: to avoid becoming obsolete, you must be open to change and willing to harness the power of new technologies. The AIvolution is here, and your willingness to adapt will determine whether you thrive or fall behind. Embrace the future, and let AI amplify your potential to unimaginable heights.

In this chapter, we will try to understand our impact in the universe, the progression of AI utilizing quantum computing and the future of humAIn to manage wealth and build connections with each other.

Our imprint in the universe

When you reflect on our lives compared to the vast timelines of the universe and the solar system, it's astounding how brief our exis-tence truly is. The universe is about 13.8 billion years old, and our solar system formed around 4.6 billion years ago. If we compress the uni-verse's entire history into one calendar year, with the Big Bang occurring on January 1st, human history only appears in the last few seconds of December 31st. An 80-year human life would equate to approximately 0.0002 seconds on this cosmic calendar. To put this into perspective, a hummingbird's wing flap takes about 0.03 seconds. Your 80-year life is only 0.67% of that sin-gle-wing flap. In the grand timeline of the universe, our entire

existence is even shorter than the blink of a hummingbird's wing. This comparison underscores how incredibly brief our time is, reminding you to cherish every moment, pursue your passions, and make meaningful connections. While short in the cosmic sense, your life is a vibrant spark that can light up the world around you.

Historically, humans lived much shorter lives. Before the 1900s, the average life expectancy was 30-35 years due to disease, malnutrition, and limited medical knowledge. Today, thanks to advancements in medicine, technology, and overall living conditions, people can live up to 100 years or more. Despite this increase, even a century is a blink in the vast expanse of the cosmos. In this brief moment we call life, all biological organisms have two fundamental goals: to pass on their genes and, for intellectual beings like us, to pass on our knowledge. This drive to reproduce ensures the survival of our species, while the transmission of knowledge helps us build on the achievements of those who came before us. This is how civilizations grow and advance—through the relentless pursuit of understanding and sharing that understanding with future generations. Your genetic code carries instructions for replication, ensuring that life continues. Simultaneously, your intellectual pursuits—scientific discoveries, technological innovations, cultural creations—are passed down, adding layers to the foundation of human progress. Each generation stands on the shoulders of the previous ones, striving to reach greater heights.

As you navigate your journey, remember that efficiency and enjoyment are not mutually exclusive. While it's essential to be productive and contribute meaningfully to this continuous flow of knowledge, it's equally important to savor the moments. Life is incredibly short, and the pressures of modern society can make you forget to pause and appreciate the beauty around you. Take a moment to reflect on your place in this grand tapestry. Feel the connection to the past, to the countless generations that have lived, loved, and learned before you. Understand that your contributions, no matter how small they may seem, are part of something much larger. Embrace the dual responsibility of passing on your genes and knowledge while finding joy in the here and now. Live with purpose but also with passion. Cherish your loved ones, pursue your passions, and remember to breathe in the wonder of existence. You are a vital link in the chain of life, and your actions today will ripple into the future.

In the context of AI, our brief existence gains even more significance. The integration of AI into our lives is a profound extension of our intellectual pursuits, amplifying our abilities and extending our reach. AI can process vast amounts of data, recognize patterns, and provide insights that enhance our understanding of the world. By embracing AI, we can magnify our intellectual contributions, ensuring that our brief time here leaves a lasting impact. To make the most of these moments, it is essential to understand and utilize AI efficiently and calmly. Accepting the reality of our limited time can inspire us to leverage AI as a tool for enhancing our productivity and creativity, allowing us to focus on what truly matters. By integrating AI into our lives thoughtfully, we can navigate the complexities of the modern world more effectively, balancing our drive to achieve with the simple pleasures that make life worth living.

In particular, the future integration of quantum computing and AI has explosive potential that may ultimately change how we think of AI as a separate entity.

Quantum AI

The fascinating world of quantum computing and its potential to transform our future is already here. Quantum computing is not just a step up from classical computing; it's a whole new paradigm. While traditional computers use bits as the smallest unit of data, represented by 0s or 1s, quantum computers use qubits.[113] These qubits can exist in multiple states at once, thanks to the principles of superposition and entanglement. This means they can process a vast amount of data simultaneously, something classical computers can't even dream of doing.

You might wonder where we stand with quantum computing today. Companies like IBM, Google, and D-Wave are leading the charge. This rapid advancement is why countries worldwide, including the United States, China, and the European Union, are investing heavily in quantum technology.

If you want to simulate the entire universe at a quantum level, we will need around 300 qubits. The IBM computer can already process more than 400 qubits.[114] For example, in drug discovery, quantum computing can analyze complex molecular structures much faster than traditional methods, potentially reducing the time to find new treatments from years to months.

AI, when integrated with quantum computing, will magnify these capabilities. Quantum computing can handle and analyze massive datasets with unprecedented speed, enhancing AI's ability to learn and make predictions. This combination could revolutionize fields like healthcare, where personalized treatments based on individual genetic profiles could become the norm.

Quantum computing is not set to replace classical computing but to complement it. Classical computers are still the best for routine tasks and data storage. Quantum computers will tackle the problems that are too complex for classical systems. This synergy will push the boundaries of what we can achieve with technology.

Currently, several countries are racing to achieve quantum supremacy. The United States has initiated the National Quantum Initiative Act to boost quantum research. China has invested billions and built the world's longest quantum communication network. The European Union's Quantum Flagship initiative is also fostering innovation in this field.

Consider how quantum computing will integrate into our daily lives as you think about the future. Just like the internet and smartphones have become essential, quantum computing combined with AI will revolutionize our interaction with technology and each other. This transformation is happening now, and the possibilities are endless.

Human integration with quantum computing will grow exponentially. By staying informed and embracing these advancements, you can be part of this exciting journey into the future of computing. This isn't a distant vision; it unfolds right before you. Quantum AI will revolutionize AI integration at an exponential level with human cognition.

Despite rapid progress, acceptance of evidence has always been a slow uptake.

Science progresses one funeral at a time

Let me explain the magic of science and the formation of "ME Thinking." This idea is rooted in the fascinating world of scientific exploration and evidence-based reasoning. Let's visualize this scenario where I take you to the past: you're exploring the ancient Mayan civilization in Mexico, where for over 10,000 years, they worshipped the Sun God. When the sun didn't shine, they believed it required a blood sacrifice to bring it back. Fast forward to today, and we see how far we've come. Our understanding of the world has evolved dramatically thanks to science.

In the last 400 years, science has transformed our world by building on previous knowledge. I recall attending a conference in Tampa, Florida, where a new topic in biology had everyone buzzing. An elderly scientist who had been researching for 52 years admitted his lifelong work was incorrect based on new evidence. The audience applauded because that's the beauty of science— it evolves with new discoveries. As scientists, we don't cling to beliefs; we rely on evidence. This constant reevaluation based on evidence is what science is all about. We formulate hypotheses, test them rigorously, and adjust our conclusions as new data emerges. This dynamic process contrasts sharply with polarized thinking, which often clings to rigid beliefs despite new evidence.

Recently, the scientific community has faced challenges, such as the retraction of nearly 10,000 papers in 2023 alone. While this might seem concerning, it highlights science's self-correcting nature. Mistakes are acknowledged and corrected, ensuring our understanding continually improves. This process doesn't mean we should abandon science for superstition; it underscores the importance of maintaining an evidence-based approach.

"ME Thinking" draws inspiration from this scientific methodology. When encountering a problem, create a null hypothesis—

the opposite of what you believe—and seek evidence to disprove it. If you can't, then your original hypothesis gains credibility. For instance, if you believe the Earth is round, your null hypothesis would be that the Earth is not round. Through experimentation, if you find no evidence to support the null hypothesis, you accept that the Earth is round.

This approach can be applied to everyday life, allowing anyone to think like a scientist without formal training. You can make informed decisions by triangulating opposite information, verifying sources, and questioning motivations. This critical thinking is crucial for passing accurate knowledge to future generations, fostering connections, and reducing the suffering caused by polarized thinking.

Our understanding of the world is constantly evolving. For example, while we often describe the Earth as round, it's an oblate spheroid. Such nuances remind us that knowledge is complex and multifaceted. Embracing this humility is essential as we integrate AI into our lives. AI is not a panacea or a threat but a tool that, depending on how we use it, can help us navigate the information age more effectively. Therefore, how you think will ultimately guide your internal AI to forage information. The power is in YOU. You must know the basics of ME thinking and mindset to overcome your mental power. The process starts with overcoming FOMO.

Embracing the moment: Overcoming FOMO

In today's fast-paced, achievement-driven world, the fear of missing out, or FOMO, has become a pervasive force. Social media platforms constantly bombard you with images of friends achieving great things, traveling to exotic locations, and seemingly living perfect lives. This relentless comparison can make you feel like you need to constantly strive for more, achieve more, and be more. However, it's crucial to understand that it's okay not to be in a perpetual state of achievement. You don't have to be the most efficient person in the world.

FOMO often drives people to set and chase goals incessantly, leading to elevated stress levels. When you are always trying to achieve something, your body releases cortisol, the stress hormone. This state of constant stress can prevent you from enjoying life's simple pleasures and can even hinder your creativity. Research has shown that creativity flows best in a relaxed and calm state. When you are relaxed, your mind is free to wander and explore new ideas. Just like the wise farmer who was content with what he had, you need to learn to relax and enjoy the moments of life.

The fear of missing out is largely a social media or peer pressure illusion, an unrealistic standard we've created. Social media has perpetuated this illusion, leading to a noticeable decline in mental health. Parents, especially those who adopt a tiger parenting style, often feel their children must achieve extraordinary success to be deemed worthy. This relentless pressure can have severe consequences. It's essential to understand that it's okay if your child doesn't go to college or doesn't achieve the dreams you envisioned for them. It's their life, their journey. Let them find their path and passion. Take a moment to observe the flap of the hummingbird and I can bet you will not be able to see one flap of the wing as it happens so quickly! That one flap is slower than your lifespan in contrast to the universe's timeline.

Consider this: in the Western world, people often live in abundance yet experience high levels of unhappiness. This paradox shows that material abundance and high achievements do not guarantee happiness. Instead, true happiness often arises from cultivating a calm state of mind, nurturing respectful relationships, and pursuing a purpose greater than ourselves.

Visualize this scenario where your child goes to the best school in the county, but two blocks away, another child is caught in a cycle of violence due to economic disadvantage. The disparity highlights a crucial point: achieving personal success while ignoring societal issues is futile. The fear of missing out drives people to focus on individual success, often at the expense of community well-being. What's the point of achieving great personal success if you are not contributing to building a kind and inclusive society?

Take a step back and reflect on your motivations. Are you pushing yourself and your children to achieve for the right reasons? Understand that true fulfillment comes from living a balanced life, appreciating the present moment and cherishing your relationships. It does not come in focusing on building personal wealth only. In Japan, "Ikigai" refers to finding purpose and meaning in everyday life. This balance leads to a fulfilling life, not just a life filled with achievements. So, encourage ME (Mindfulness and Effortless) thinking as we discussed in chapter 9.

Parents need to relax and allow their children to explore their interests and passions. Letting go of rigid expectations and the constant push for success can lead to happier, more well-rounded individuals. Encourage your children to follow their goals, even if they don't align with traditional measures of success. For example, if your child is passionate about art, support their creative endeavors rather than push them towards a high-paying but unfulfilling career.

Society benefits when individuals are not just successful but also content and kind. A study by the American Psychological Association found that children who grow up in supportive environments, where they are allowed to pursue their interests, tend to have better mental health and are more likely to contribute positively to their communities.[115]

Instead of focusing solely on academic or professional success, prioritize fostering empathy, creativity, and kindness in your children.

Finally, consider the impact of social media on your perception of success. The curated lives you see online are not the full picture. People tend to share their highlights, not their struggles. By constantly comparing yourself to these idealized versions of life, you feed into the FOMO cycle. Instead, take regular breaks from social media from concepts of our chapter, Digital Detox and focus on real-life connections and experiences. Practice gratitude and Mindfulness to appreciate what you have rather than what you lack. The ME techniques can support the next generation in curating knowledge and increasing their chances of efficiently filtering out crucial information. People often ask me if it is human versus AI and if AI will extinct all human races. I always answer it will be "human VS humAIn".

Human Vs. humAIn

In the future, it won't be AI versus humans. It'll be human versus humAIn. Yes, you heard that right! The difference between those who embrace AI and those who don't will be like night and day. Those who harness the power of AI will have an edge in everything. Those who integrate AI into their lives and work will undoubtedly outpace those who ignore or resist this incredible technology.

Historically, early adopters of technology have always gained a significant advantage. For example, the McKinsey Global

Institute reports that early adopters of AI technologies see profit margins over 20% higher than those who are slow to adopt.[116]

This isn't just about having more information; it's about using it effectively to stay ahead in a world that's evolving faster than ever. AI brings together many disciplines that we may not have previously considered relevant to each other. To truly leverage AI, you need to have insights from neuroscience, data science, computer engineering, and ethics. Throughout this book, we've explored how these fields intersect to shape the development and implementation of AI. Understanding these intersections helps us use AI more effectively and responsibly, making it a powerful tool for personal and professional growth.

Think of AI as a tool, much like your word processor or smartphone. It doesn't replace human intellect; it enhances it. Just as word processors have made writing and editing more efficient, AI can boost our cognitive abilities. It processes vast amounts of data, identifies patterns, and provides insights our brains might miss. This doesn't diminish your intelligence; it complements it, making you smarter, more efficient, and more informed. Howard Gardner's theory of multiple intelligences identifies different types of intelligence that play unique roles in human cognition: linguistic, logical-mathematical, spatial, bodily-kinesthetic, musical, interpersonal, intrapersonal, naturalistic, existential, and moral.[117] Each type is essential in various contexts. For instance, linguistic intelligence is crucial for effective communication, while logical-mathematical intelligence is key for problem-solving. AI can enhance these intelligences by providing tools and insights that augment our natural abilities.

We will soon have another intelligence adding to our humAIn

intellect: internal AI intelligence. This personalized AI understands you intimately, learns from your interactions, and helps you make decisions aligned with your goals and values. Picture this: each of us has our own augmented AI, much like how we each have a smartphone today. The critical difference is that this internal AI will be tailored to our needs and preferences, reflecting our unique thought processes and values. It will be integral to our daily lives, guiding and optimizing our routines.

Internal AI will help you navigate the overwhelming flow of information, filtering out noise and highlighting what's truly important. It will support your mental and emotional well-being, offering insights into your behavior and suggesting ways to improve your life. Imagine an AI that manages your schedule, optimizes your work processes, and suggests ways to enhance your health and well-being. This isn't science fiction; it's the reality we're moving towards.

In this new era, the integration of AI into our lives requires a mindful approach. AI should not replace human judgment or creativity; it should augment them. By integrating AI into your daily routine, you can make better decisions, solve complex problems, and unlock new levels of creativity. However, we must also ensure that AI is used ethically and responsibly. As you navigate this AI revolution, remember that your choices and openness to new possibilities will shape the future. The future of AI is about collaboration and augmentation, not replacement.

We are now at the dawn of the AI revolution. The pattern is clear: those who leverage AI will advance, while those who resist will fall behind. This revolution is not just about technology; it's about enhancing human capabilities and potential. Embrace this change with a clear understanding of the ethical and practical implications.

Our lives, in the universe's grand scheme, are fleeting moments. Even living to 100 years is a blink in cosmic time. In this brief span, we have two primary goals: passing on our genes and knowledge. We can live longer and more fulfilling lives with modern technology and medicine, but longevity alone is not enough. We must also strive for quality and meaning in our existence. This is where AI comes into play, helping us achieve our goals more efficiently and giving us more time to enjoy the moments that truly matter.

You don't need to carry the world's weight on your shoulders. With AI, you can streamline tasks, make informed decisions, and focus on what you love. This revolution is about enhancing your life, not overwhelming it. By integrating AI mindfully, you can navigate the complexities of the modern information age effectively, enjoying a balanced and fulfilling life. The techniques in this book will guide you to adapt to AI and algorithms for improving every aspect of your life. Specially, we need to start thinking of one humAIn species rather than competing with our kind. Maybe we can emerge as one species and one human feder-ation soon by eliminating the personal wealth dilemma.

End of personal wealth

Think back to the Stone Age. During that time, the concepts of personal wealth, nuclear families, and ownership were nonexistent. Life was communal, with people working together for the common good. The community raised children, and there was no notion of "my wife," "my husband," or "my children." Everything was shared, and the sense of unity was strong.

Fast forward to our modern world, where individualism and personal wealth dominate. However, our lives are fleeting when contrasted with the expanse of the universe. Our time here is incredibly short. What if we shifted our perspective back towards community and unity, not just within families or societies, but as one global species—humAIn?

The world was divided in the past by borders, cultures, and ideologies. Today, we're moving towards greater unification. A future without countries where we focus on creating wealth and opportunities for all humans is within reach. This dream of a single human federation, where resources are shared equitably, is possible, and AI holds the key to making this vision a reality.

AI can potentially eliminate the biases plaguing our current systems of government, regulation, and law—systems created by humans and, therefore, inherently biased. As we evolve, we must acknowledge that our cognitive capacities are limited by the environments we grow up in. A child raised in Australia will see the world differently from one raised in the United States or Madagascar. Our beliefs and actions are shaped by the stories we're told and the environments we inhabit.

At our core, we all share the same consciousness and fleeting moments in time and space. It's essential not to let ego drive us toward conflict and division. History is littered with the consequences of such egotism—countless lives lost because of a king's pride or a nation's desire for dominance. AI can help us transcend these limitations. With its potential for unbiased decision-making, AI can support a fair and unified future. The fearmongering around AI often stems from a desire to control and profit, turning people into products. But if we learn to trust AI, we can create a world where personal wealth is a thing of the past and competition is about bettering ourselves rather than outdoing each other. By embracing AI, we can stop competing against each other and start striving for the collective betterment of humanity. This is our chance to shift from individualism to unity, from personal wealth to shared prosperity, and AI is the tool that can help us achieve this beautiful future.

Home beyond home

You start soaking up information like a sponge from the moment you're born. As humans, we're unique among mammals because our journey begins with a hefty dose of vulnerability. We can't even walk right away, so we naturally rely on our parents and immediate environment for everything. This early trust shapes our beliefs and perceptions, and we grow up thinking that the world our caregivers introduce us to is the ultimate truth.

But here's the kicker: as you grow older and start exploring the world on your own, you realize there's a lot more out there than what your childhood bubble showed you. The exciting and sometimes scary part is that your potential can feel boxed in by societal norms, family expectations, cultural dictates, and even deeply held personal beliefs. It's like living in a cozy but sometimes stifling cage.

Enter the age of social media. It's a double-edged sword. On one side, it's a breeding ground for misinformation. You know those dubious articles your aunt keeps sharing? Yeah, those. On the flip side, social media can be an amazing tool for finding your tribe. Suddenly, you're not alone in your quirky interests or niche passions. There's a whole community out there that gets you.

Now, let's talk about AI. Think of a best friend who always knows what you need—no, not because they're psychic, but because they've been paying attention. Your AI can become that friend. It learns from the information you engage with and helps you find more of what you're looking for. The next phase of human evolution could be creating a sense of belonging that transcends physical boundaries. Whether you move to a new country or, who knows, a new planet someday, your AI friend is right there with you, making the new place feel like home.

People often get boxed into their current environment because it's what they know. But with AI, you have the chance to break free and find the place where you truly belong. It's an exciting time because "home" can now be a concept that travels with you, shaped by your AI companion.

Of course, you need to build a good relationship with your AI. Treat it like a pet—train it well, interact with it positively, and it will become your loyal companion. But remember, AI is only as good as the information you feed it. If you fill it with negativity and misinformation, that's what it will give back. So, be open-minded and teach your AI to help you navigate the world effectively.

The world is your oyster, and your AI is the pearl inside. It's up to you to make the most of it. We need kind and empathetic humAIn for the next step of AIvolution.

Creating kind humAIn

In today's age of information and AI, we are more connected than ever, sharing ideas and experiences with millions of people through social media. This interconnectedness has the potential to create a collective intelligence, a "bigger mind," where we can learn from each other and grow together. However, this powerful tool comes with a responsibility. If we propagate hate, it will permeate our shared spaces, leading to a divisive and hostile world. Instead, we must embrace kindness and understanding, fostering a supportive and inclusive environment.

The ongoing AIvolution, the integration of AI into our daily lives, is not something to fear but to welcome with open arms. AI can enhance our efficiency, productivity, and overall capabilities, making us more amazing than ever. But this can only happen if we approach this evolution with Mindfulness and positivity. Without a proper understanding and a mindful approach, we risk becoming slaves to the algorithms of social media companies, driven by likes and shares rather than genuine human connection.

As we enter this new era, being kind to one another is crucial. Our interactions online can have a profound impact on real-world behaviors and attitudes. We contribute to a more harmonious and enlightened society by sharing positive and constructive content. This AIvolution is just beginning, and we have the opportunity to shape it into a force for good. Embrace this change, understand the power of your actions and words, and become a beacon of positivity in the digital age. Together, we can navigate this transformation and emerge as the best versions of ourselves, fully integrated with AI yet grounded in human values.

Ever felt that instant, inexplicable connection with someone new, like your souls were destined to meet? That's the kind of incredible journey we're about to explore together – a tale where lives across centuries intertwine in ways that defy expectation. Love, betrayal, the quest for freedom – it's all here, reminding us that despite the vast distances and eras, the human spirit is endlessly interconnected. And guess what? AI is creating connections on a global scale, just like that. It's your new best friend, linking you with brilliant minds worldwide, igniting collaborations you never imagined possible. This isn't just about technology; it's about harnessing the incredible power of human ingenuity, supercharged by AI.

But here's where it gets exciting: your choices about every decision and interaction send ripples through time. Just like the characters in our stories in this book, the choices you make with AI – how you develop and use it – will shape not only your future but also the future of generations to come. It's a huge responsibility but also an exhilarating opportunity.

And let's not forget the power of belief. In the stories mentioned in this book, characters hold onto faith, hope, and love as their compass in a chaotic world. As AI challenges your assumptions and makes you question everything, these core human values will guide you through uncharted territory. So, what's the big message? It's simple yet profound: You're connected to everyone, and AI is amplifying that connection in unprecedented ways. This isn't about fearing the future but embracing it with open arms and a mindful heart. Let's choose wisely, collaborate passionately, and use AI to build a future that honors our shared humanity.

Remember, YOU are the humAIn at the heart of this AIvolution. Your choices matter, your voice matters, and your potential is limitless. So, be brave and explore this exciting new world together, hand in hand with AI, and create a future that celebrates the very best of what it means to be the next emerging humAIn.

epilogue

TOWARDS AIVOLUTION

Hard times create strong men. Strong men create good times.
Good times create weak men. Weak men create hard times.
- G Michale Hopf

You might have picked up this book expecting a deep dive into the technical intricacies of AI. However, any meaningful journey begins within yourself—starting with your mindset and the misinformation surrounding AI in today's digital age. There's a lot of noise out there, and many influential voices either stoke fear about AI or advise against its integration into our lives. The purpose of this book is to help you see through that fog and realize that AI can be a powerful ally, even if you have little or no technical knowledge.

First and foremost, understand that AI is not your enemy. By now, hopefully, your fears about AI have diminished. Think of AI as a tool, much like a Swiss Army knife, capable of a multitude of tasks that can enhance your life and work. My expertise in AI and healthcare has driven me to help others harness the power of AI for the greater good. That's the motivation behind this book.

You have incredible potential. Whether you are a student contemplating college, a college student considering higher studies, or a professional looking to pivot into technology or education, we are here to support you. Our team of educators has already assisted thousands of people, and you can be one of them. Your limitless potential can be fully realized with the aid of AI. Becoming a human enhanced by AI is the best possible path because you already possess the most important trait: humanity. AI is here to augment your capabilities, not strip away your kindness and empathy.

It's a common misconception that AI and robots lack empathy and kindness. But what truly defines kindness? It's the capacity to show empathy, understand others' mistakes, and provide growth opportunities. AI can help us achieve this by supporting those who aren't yet super-efficient members of society and giving them the tools to improve. In many ways, AI can enhance our ability to be kind and empathetic by providing personalized assistance and support.

As we stand on the threshold of the early 21st century, we find ourselves amid global unrest, economic uncertainty, and societal upheaval. This turmoil isn't new; it's part of an economic and social cycle that has been unfolding throughout human history. Civilizations have risen and fallen: the Romans and Greeks once held immense power, only to crumble under their own weight. The British Empire reigned supreme until it declined, and China, too, has seen its share of ascendancy and downfall.

This pattern is often driven by generational shifts. One generation toils to build wealth and stability, the next enjoy the fruits of that labor, and the subsequent generation often faces diminished opportunities and mounting frustrations. After World War II, the world experienced a period of prosperity and relative peace. The generation that emerged from the ashes of war worked tirelessly to rebuild and innovate, creating a legacy of economic growth and technological advancement. Fast forward to today, and we see the echoes of history in the struggles of Generation Z. Born into a world of rapid technological change and complex global challenges, they face economic constraints and political pressures that threaten their future. This has led to widespread disillusionment and unrest. We are witnessing the rise of protests and the fall of governments as young people demand change and accountability.

What's different now, compared to the past 1,500 years, is the unprecedented connectivity of our world. A protest in Nairobi can be witnessed in real-time in Madagascar and Mongolia, thanks to social media platforms. This global interconnection empowers individuals and movements, enabling them to share ideas and unite for common causes. However, it also brings challenges. Disinformation and misinformation spread rapidly, fueled by algorithms that can deepen divisions and amplify conflicts. AI plays a crucial role in shaping our perceptions and interactions. The algorithms behind our digital platforms are not neutral; they reflect and reinforce our biases and behaviors. How we design and deploy AI will significantly impact our future. It is imperative that we approach this with care, striving to create systems that promote understanding and collaboration rather than division.

In this critical moment, your actions matter. The choices you make and the values you uphold can shape the world for generations to come. Treat each other with kindness, empathy, and respect. Advocate for systems that prioritize human well-being over profit. Demand transparency and accountability from those who wield technological power. The next generation of humanity—what we might call "humAIn"—must be defined by its commitment to justice, equity, and inclusivity. As global citizens, we have the opportunity and responsibility to create a world where technology serves us all, not just a privileged few. Embrace this challenge with enthusiasm and conviction. Your voice and your actions can help shape a brighter, more equitable future for all.

In the words of Martin Luther King Jr., "The arc of the moral universe is long, but it bends toward justice."[118] Let's ensure that our actions help it bend in the right direction. Together, we can build a world where every person has the opportunity to thrive, where technology is a force for good, and where future generations inherit a legacy of compassion and progress. AI offers immense potential to enhance your life and career. The key is to approach it with an open mind and a willingness to learn. This book is your guide to navigating the AI landscape, dispelling fears, and embracing the opportunities that lie ahead. Follow us, engage with our content, and let's build a better future where technology and humanity coexist and thrive.

Engage with the content of this book and stay connected with us through social media and future endeavors. Don't miss out on the wave of opportunity that AI presents. The digital landscape is rife with misinformation; without the right guidance, it's easy to get lost. By following latest AI news and the community, you can stay informed, make better decisions and leverage AI to its fullest potential. Think of this as your opportunity to connect with a vibrant community of like-minded individuals excited about AI's future. Don't be left behind. The rapid pace of technological advancement means that those who are informed and prepared will have a significant advantage. By engaging with the resources and support we offer, you can ensure that you are one of the pioneers, not just a spectator, in the AI revolution. So, take the first step. Participate in our community and start harnessing the power of AI today. Your future self will thank you for becoming the productive and kind humAIn.

additional
READING

Lets do a quick exercise using ME thinking.

Exercise: AI will take away your job.

*How to Use ME-Thinking to Combat the Fear of AI
Taking Over Jobs*

Let's face it, the idea that AI will take over all jobs can be scary. But with ME-thinking—Mindful and Effortless thinking—you can manage this fear effectively. Let's walk through an exercise to show you how.

Step-by-Step Guide to M-Thinking

First, engage your M-thinking. This is your mindful, analytical side that helps you process information deeply and critically.

Step M. Metacognition (Thinking about Thinking)

Start by reflecting on your thought process. Ask yourself, "Why am I accepting this fear as true? What biases might I have?" Recognizing these can help you think more critically and less emotionally about the situation.

Step I. Information Source Motivation

Consider the motivation behind the sources you encounter. Are they credible and reputable? For instance, an article from a respected industry expert discussing AI's role in your field is more reliable than a random blog. Understand if there is any financial gain for the source spreading fear about AI.

Step N. Navigate Opposite Information

Look for information that presents an opposing viewpoint. If you're a nurse, search for articles explaining how AI can assist but not replace the human element in nursing. For example, AI can handle administrative tasks, but patient care, empathy, and decision-making still require human touch. Use studies, expert opinions, and real-life testimonials to get a comprehensive view.

Step D. Determine the Worst That Can Happen

Ask yourself, "What's the worst that can happen?" Often, the worst-case scenario is less severe than imagined. For instance, even if AI changes your job, it might lead to new opportunities and roles that didn't exist before.

Step S. Statistical Thinking

Numbers provide clarity. Look at statistics about AI's impact on jobs. Reports often show that while some jobs evolve, new opportunities arise. For example, studies indicate that AI will create more jobs than it eliminates, particularly in sectors requiring human creativity and emotional intelligence.

Step E. Evaluate Higher Purpose Beyond Yourself

Think about your higher purpose in your job. For example, as a nurse, your purpose is to provide care and empathy—qualities AI cannot replicate. This purpose gives you clarity and motivation to adapt and grow.

Step T. Track Your Endpoint

Start with your end goal in mind. If your goal is to remain relevant in your job, identify what skills and knowledge you need to stay indispensable. For example, if AI is enhancing certain tasks, learn how to use AI tools to improve patient care or efficiency in your role.

Now, lets look at how you can use your autopilot of E thinking in this scenario.

Step-by-Step Guide to E (Effortless)-Thinking

Once you've engaged in M-thinking, you can move to E-thinking—effortless thinking—which allows you to process information smoothly and efficiently.

Step C: Collect and Internalize Evidence

After finding solid evidence that AI won't take your job, internalize this information. Recognize that while AI might change how some tasks are performed, it won't replace the essential human aspects of your job. Figure out where automation can save money for business and it will help you to predict which areas of your job may be automated.

Step A: Activate Effortless Mode

Allow your brain to shift to a more relaxed, automatic mode when you encounter similar information in the future. You've done the hard work of researching and analyzing, so trust your findings. Don't panic on sudden video posts or blogs. Remember, at the end those articles were written by one person.

Step L: Lodge Positive Beliefs

Each time you see a fear-inducing headline about AI and jobs, remind yourself of the evidence you've gathered. This reinforcement will help maintain a balanced perspective. Keep track of your positive flow of information. Engage in activities that bring you joy and fulfillment. If you're passionate about nursing, focus on patient care and continuous learning. This immersion can lead to a state of flow, where you're fully absorbed and time flies, blending M and E thinking.

Step M: Maintain a Balanced Digital Diet

Regularly review your social media and news feeds. Ensure that the information you consume is balanced and not just sensationalist content designed to provoke fear. This practice helps keep your E-thinking mode healthy and productive.

ACKNOWLEDGMENTS

First and foremost, I would like to express my deepest gratitude to my parents, Nurul and Rokeya, whose unwavering love and support laid the foundation for the person I am today. My childhood, filled with warmth, curiosity, and the wonders of imagination, was a gift that continues to inspire me. It was in those early years, immersed in the worlds of Star Wars and Star Trek, that the seeds of my fascination with science and technology were planted. My dream of becoming a scientist, driven by a desire to create and discover, was born out of the boundless encouragement and opportunities my parents provided. Their belief in me has been a constant source of strength, guiding me through the journey that led to this book and my research in artificial intelligence.

To my brothers, Sunny and Bonny, I owe a debt of gratitude that words can scarcely express. Your steadfast support, encouragement, and belief in my dreams have been the bedrock upon which I've built my career and life. You have always stood by me, believing in my vision even when the path was uncertain. Your unwavering confidence in my abilities has given me the courage to pursue my passions, and for that, I am profoundly grateful.

This book is also a testament to the inspiration I have drawn from the incredible students I have had the privilege to teach. It is often said that teachers shape the minds of their students, but in my experience, it is the students who have shaped mine. Your curiosity, energy, and fresh perspectives have not only broadened my understanding but have also rekindled my passion for learning. As you explored the fields of AI and machine learning, you challenged me to see the world through your eyes, to question more deeply, and to think more creatively. Your influence has been a driving force behind the recognition I've received in teaching and research, and I am forever indebted to you.

I would also like to extend my heartfelt thanks to my extended family, whose support has been a constant source of comfort and strength. Life is full of ups and downs, and it is the love and encouragement of family that helps us navigate the most challenging moments. My sister-in-law, Sanjida, and my nephews, Areeb and Neil, you have brought joy, inspiration, and a sense of wonder into my life. Your youthful curiosity and fearless questioning remind me of the importance of staying curious and open-minded. You challenge me to think differently and inspire me to continue pushing the boundaries of knowledge. Your presence in my life has been a blessing, and I am deeply grateful for your love and support. I am very indebted to Erena, Niloy, Jemi and Mamun for inspiriting to make this book come alive.

A special acknowledgment goes to Maryann, whose unwavering support and belief in the importance of this book have been instrumental in bringing it to life. In a world where AI is often met with fear and uncertainty, Maryann saw the need for a different narrative—a narrative of hope, inspiration, and possibility.

Her dedication to this project, from the earliest stages of writing to the final edits, has been invaluable. Marianne, your belief in this book and your tireless efforts to help shape it into what it is today have made all the difference. Thank you for being a true partner in this journey.

Finally, to my son, Neo—this book is for you. You are my greatest inspiration, and it is my hope that as you grow, you will come to understand the incredible beauty and complexity of the world we live in. Life is not about controlling every moment but about experiencing them fully, learning from them, and passing on the knowledge we gain to future generations. I hope that this book will offer you insights into some of life's deeper philosophies and that it will inspire you to approach life with curiosity, courage, and an open heart. May you always find joy in discovery and fulfillment in the pursuit of knowledge.

In closing, I want to acknowledge all those who have walked this journey with me, whether directly or indirectly. Every word in this book is a reflection of the collective wisdom, love, and support I have received from so many. This book is not just a product of my efforts but a culmination of the influences, teachings, and inspirations of countless individuals. To everyone who has played a part in this journey—thank you. Your contributions have made this book, and my journey, what they are today. Together, we are shaping the future, one thought, one idea, and one innovation at a time.

About the Author

DR. DON ROOSAN is a global leader in artificial intelligence integration and a passionate advocate for using AI responsibly to improve our world. An internationally recognized scientist and researcher, he has spent his career guiding individuals and businesses through the complexities of AI, helping them foster innovation while maintaining ethical integrity. Known for his engaging style and visionary perspective, Dr. Roosan inspires people to see AI as a tool for enhancing human potential rather than replacing it.

As a faculty member at the School of Engineering and Computational Sciences at Merrimack College in Boston, Massachusetts, Dr. Roosan is at the forefront of shaping the future of AI in a way that benefits society. His mission is clear: to ensure AI is a force for good, driving progress and a brighter future for all.

Notes

Chapter 1

1. Technology Org. "Artificial Intelligence and Unemployment: The Good & Bad." *Technology Org*, September 17, 2022. https://www.technology.org/2022/09/17/the-impact-of-artificial-intelligence-on-unemployment/.

2. Blouin, L. "AI's Mysterious 'Black Box' Problem, Explained." *University of Michigan-Dearborn*, March 6, 2023. https://umdearborn.edu/news/ais-mysterious-black-box-problem-explained.

3. Budiu, Raluca. "Information Foraging: A Theory of How People Navigate on the Web." *Nielsen Norman Group*, November 10, 2019. https://www.nngroup.com/articles/information-foraging/.

4. Common Good Ventures. "Common Good Ventures." Accessed June 21, 2024. https://commongoodventures.org/.

5. Coughlin, T. "175 Zettabytes By 2025." *Forbes*. Accessed November 27, 2018. https://www.forbes.com/sites/tomcoughlin/2018/11/27/175-zettabytes-by-2025/.

6. Coursera. "Coursera | Online Courses & Credentials by Top Educators. Join for Free." *Coursera*, 2018. https://www.coursera.org/.

7. Economic Policy Institute. "Education Inequalities at the School Starting Gate: Gaps, Trends, and Strategies to Address Them." *Economic Policy Institute*. Accessed June 2024. https://www.epi.org/publication/education-inequalities-at-the-school-starting-gate/.

8. Khan Academy. *Khan Academy*. Accessed 2024. https://www.khanacademy.org/.

9. "New Algorithm Charter a World-First." *The Beehive*. Accessed 2024. https://www.beehive.govt.nz/release/new-algorithm-charter-world-first.

10. Michael. "Modular Skills Trainer." *Knowles Training Institute*, July 26, 2023. https://knowlesti.sg/modular-skills-trainer/.

11. Soy, S. *Research in Multidisciplinary Subjects (Volume-1)*. The Hill Publication, 2023.

12. Thomas, D. "What is the Digital Age?" *Ventiv Technologies*, August 22, 2019. https://www.ventivtech.com/blog/what-is-the-digital-age.

13. "University of Richmond Artificial Intelligence Boot Camp Overview." *University of Richmond Boot Camps*. Accessed June 22, 2024. https://bootcamps.richmond.edu/artificial-intelligence/.

Chapter 2

14. Fridman, L. "Wojciech Zaremba: OpenAI Codex, GPT-3, Robotics, and the Future of AI." *Lex Fridman Podcast #215*, YouTube, 2021. https://www.youtube.com/watch?v=U5OD8MjYnOM.

15. Waymo. *Waymo*, 2023. https://waymo.com/.

16. "Self-Driving Car Statistics." *NST Law*. Accessed June 2024. https://www.nstlaw.com/autonomous-vehicle-statistics/.

17. National Institute of Mental Health. "Technology and the Future of Mental Health Treatment." *NIMH*, 2019. https://www.nimh.nih.gov/health/topics/technology-and-the-future-of-mental-health-treatment.

18. IBM. "What Is Artificial Intelligence (AI)?" *IBM*, 2024. https://www.ibm.com/topics/artificial-intelligence.

19. "Deep Learning in ArcGIS." *Esri Indonesia*. Accessed June 23, 2024. https://esriindonesia.co.id/store/deep-learning-arcgis.

20. Schrödinger, Erwin. "Die gegenwärtige Situation in der Quantenmechanik." *Naturwissenschaften* 23, no. 48 (1935): 807-812.

21. "Learning Interpretability Tool." *Pair-Code GitHub*. Accessed June 25, 2024. https://pair-code.github.io/lit/.

22. Chen, H., Dong, G., and Li, K. "Overview on Brain Function Enhancement of Internet Addicts through Exercise Intervention: Based on Reward-Execution-Decision Cycle." *Frontiers in Psychiatry* 14 (2023). https://doi.org/10.3389/fpsyt.2023.1094583.

23. EqualOcean. "Report: The Internet Penetration Rate of Minors in China Is 94.9%, and the Trend of Low Age Internet Access Is Increasing Year by Year." *EqualOcean*, July 20, 2021. https://equalocean.com/briefing/20210720230061441

24. BlackDoctor.org. "Why Hypertension and Heart Disease Hits Black Americans Harder." *BlackDoctor.org*, Accessed September 10, 2024. https://blackdoctor.org/why-hypertension-and-heart-disease-hits-black-americans-harder/.

25. "What-If Tool." *Pair-Code GitHub*. Accessed June 2024. https://pair-code.github.io/what-if-tool/.

Chapter 3

26. APA PsycNet. "Inattentional Blindness and its Relevance to Teaching Forensic Accounting and Auditing." *Journal of Accounting Education* 29, no. 1 (2011): 37–49. https://psycnet.apa.org/doiLanding?doi=10.1037%2F0033-295X.106.4.643.

27. Library of Congress. "Frequently Asked Questions." *Digital Collections Management Compendium.* Last modified August 19, 2019. https://www.loc.gov/programs/digital-collections-management/about-this-program/frequently-asked-questions/."Brain Battery." *Knowing Neurons*, December 14, 2012. https://knowingneurons.com/blog/2012/12/14/brain-battery/.

28. American Psychological Association. "Speaking of Psychology: Why We Get Conned, with Daniel Simons, PhD, and Christopher Chabris, PhD." *YouTube*, July 19, 2023. https://youtu.be/NVdwQiiY6_0?si=m1MRJw1SMKEKOO8x.

29. Cleveland Clinic. "Brain Facts." *Healthy Brains by Cleveland Clinic*, 2015. https://healthybrains.org/brain-facts/.

30. "Information Foraging Theory." *Interaction Design Foundation.* Accessed June 2024. https://www.interaction-design.org/literature/book/the-glossary-of-human-computer-interaction/information-foraging-theory.

31.   Rowden, A. "What Is Synaptic Pruning?" *Medical News Today*, July 26, 2023. https://www.medicalnewstoday.com/articles/synaptic-pruning.

32.   Hahamy, A., Dubossarsky, H., and Behrens, T. E. J. "The Human Brain Reactivates Context-Specific Past Information at Event Boundaries of Naturalistic Experiences." *Nature Neuroscience* 26, no. 6 (2023): 1080–89. https://doi.org/10.1038/s41593-023-01331-6.

33.   Inglis, F. M., and Moghaddam, B. "Dopaminergic Innervation of the Amygdala Is Highly Responsive to Stress." *Journal of Neurochemistry* 72, no. 3 (1999): 1088–94. https://doi. org/10.1046/j.1471-4159.1999.0721088.x.

34.   Budiu, Raluca. "Information Scent: How Users Decide Where to Go Next." *Nielsen Norman Group*, February 2, 2020. https://www.nngroup. com/articles/information-scent/.

35.   Tversky, A., and Kahneman, D. "Judgment under Uncertainty: Heuristics and Biases." *Science* 185, no. 4157 (1974): 1124–31. https://www2. psych.ubc.ca/~schaller/Psyc590Readings/TverskyKahneman1974.pdf.

36.   Treffert, D. A. "The Savant Syndrome: An Extraordinary Condition. A Synopsis: Past, Present, Future." *Philosophical Transactions of the Royal Society B: Biological Sciences* 364, no. 1522 (2009): 1351–57. https://doi. org/10.1098/rstb.2008.0326.

37.   "Journal of Economic Psychology." *ScienceDirect*. Accessed June 2024. https://www.sciencedirect.com/journal/ journal-of-economic-psychology.

38.   "The Psychology of Pricing a Real Estate Property." *Inman*. Accessed June 2024. https://www.inman.com/2015/04/17/ the-psychology-of-pricing-a-real-estate-property/.

39.   Leadspace Team. "Marketing Effectiveness: What It Is and 4 Ways to Measure It." *Leadspace*, July 5, 2019. https://www.leadspace.com/blog/ marketing-effectiveness/.

40.   Shadlen, Michael N., and Kiani, R. "Decision Making as a Window on Cognition." *Neuron* 80, no. 3 (2013): 791–806. https://doi. org/10.1016/j.neuron.2013.10.047.

41.   Google Research. "BERT: Pre-training of Deep Bidirectional Transformers for Language Understanding." Accessed June 2024. https://research.google/pubs/bert-pre-training-of-deep-bidirectional-transformers-for-language-understanding/.

Chapter 4

42. "All About Netflix Artificial Intelligence: The Truth Behind Personalized Content." *Litslink*, March 21, 2024. https://litslink.com/blog/all-about-netflix-artificial-intelligence-the-truth-behind-personalized-content.

43. Pfizer. "Our Science: Pharmaceutical Development." *Pfizer*, 2023. https://www.pfizer.com/science.

44. Anthem, Inc. *IBM Case Study*. Accessed June 2024. https://www.ibm.com/case-studies/anthem.

45. Change, S. "Knowledge Nuggets: Exponential Organizations." *HelloStepChange*. Accessed June 2024. https://blog.hellostepchange.com/blog/knowledge-nuggets-exponential-organizations.

46. Ismail, S. *Exponential Organizations*. Diversion Books, 2014.

47. "OpenExO." *OpenExO*. Accessed June 2024. https://openexo.com/.

48. "Paxton AI Legal Research." *Paxton*. Accessed June 2024. https://www.paxton.ai/research.

49. "Peter Thiel." *Forbes*. Accessed June 2024. https://www.forbes.com/profile/peter-thiel/.

50. Farrow, E. "Determining the Human to AI Workforce Ratio – Exploring Future Organisational Scenarios." *Technology in Society* 68 (2022): 101879. https://doi.org/10.1016/j.techsoc.2022.101879.

51. Taylor, T. "6 Differences Between a Solopreneur and Entrepreneur." *HubSpot*. Accessed June 2024. https://blog.hubspot.com/sales/solopreneur.

52. "Artificial Intelligence (AI) Services & Solutions." *Accenture*. Accessed June 26, 2024. https://www.accenture.com/us-en/services/data-ai.

53. Barber, O. "Artificial Intelligence in Decision Making - Big Overview from InData Labs." *InData Labs*, December 2, 2021. https://indatalabs.com/blog/artificial-intelligence-decision-making.

Chapter 5

54. Perna, M. C. "No More Teachers: The Epic Crisis Facing Education in 2024." *Forbes*, January 3, 2024. https://www.forbes.com/sites/markcperna/2024/01/03/no-more-teachers-the-epic-crisis-facing-education-in-2024/.

55. "How Are Schools in Other Countries Different from America?" *Uni Hanoi*, February 8, 2023. https://articles.unishanoi.org/how-are-schools-in-other-countries-different-from-america/.

56. "Teach from Your Best Self: Teacher-Centered Professional Development." *Teach from Your Best Self.* Accessed July 3, 2024. https://www.teachfromyourbestself.org/.

57. Desilver, D. "U.S. Students' Academic Achievement Still Lags That of Their Peers in Many Other Countries." *Pew Research Center*, February 15, 2017. https://www.pewresearch.org/short-reads/2017/02/15/u-s-students-internationally-math-science/.

58. Tucker, M. "Why Other Countries Keep Outperforming Us in Education." *Education Week*, May 13, 2021. https://www.edweek.org/policy-politics/opinion-why-other-countries-keep-outperforming-us-in-education-and-how-to-catch-up/2021/05.

59. "Common Core State Standards Initiative." *Common Core.* Accessed June 2024. https://www.thecorestandards.org/read-the-standards/.

60. National Science Board. "Recent Trends in Federal Support for U.S. R&D." *National Science Foundation.* Accessed June 2024. https://www.nsf.gov/nsb/news/news_summ.jsp?cntn_id=303449.

61. Errington, Timothy M., et al. "Challenges in Assessing the Reproducibility of Preclinical Research in Cancer Biology." *Nature* 600, no. 7888 (2021): 359-360. https://doi.org/10.1038/d41586-021-03736-4.

62. Van Noorden, Richard. "More than 10,000 Research Papers Were Retracted in 2023 — A New Record." *Nature*, December 7, 2023. https://www.nature.com/articles/d41586-023-03974-8.

63. "Impacts of Federal R&D Investment on the US Economy." *Breakthrough Energy.* Accessed July 3, 2024. https://www.breakthroughenergy.org/newsroom/reports/impacts-of-federal-rd-investment-on-the-us-economy/.

64. Craig, R. "AI Will Shrink the University." *Forbes.* Accessed July 4, 2024. https://www.forbes.com/sites/ryancraig/2024/06/28/ai-will-shrink-the-university/.

65. zSpace. "zSpace | Virtual Reality for Education." Accessed September 10, 2024. https://zspace.com/.

66. Csikszentmihalyi, Mihaly. *Flow: The Psychology of Optimal Experience.* New York: Harper & Row, 1990.

67. "NAEP State Profiles." *Nation's Report Card*. Accessed June 2024. https://www.nationsreportcard.gov/profiles/stateprofile.

68. Jennings, J. "AI in Education: Keeping Humans in the Loop." *ESpark*, October 12, 2023. https://www.esparklearning.com/blog/ai-in-education-humans-in-the-loop/.

69. Jern, K. "Can AI Replace Teachers? We Finally Know the Answer." *ESpark*, January 19, 2024. https://www.esparklearning.com/blog/can-ai-replace-teachers-we-finally-know-the-answer/.

Chapter 6

70. Dimock, M., and Wike, R. "America Is Exceptional in the Nature of Its Political Divide." *Pew Research Center*, November 13, 2020. https://www.pewresearch.org/short-reads/2020/11/13/america-is-exceptional-in-the-nature-of-its-political-divide/.

71. "Weaponized Narrative Is the New Battlespace." *Defense One*. Accessed July 2024. https://www.defenseone.com/ideas/2017/01/weaponized-narrative-new-battlespace/134284/.

72. Cruz, H. D. "Brexit: A Detailed Summary of What the Hell Is Going On, for My American Friends." *Medium*, July 16, 2018. https://helenldecruz.medium.com/brexit-a-detailed-summary-of-what-the-hell-is-going-on-for-my-american-friends-61c92b04d9f7.

73. "Brexit Betrayed: How AI and Social Media Hijacked Democracy." *LinkedIn*. Accessed July 11, 2024. https://www.linkedin.com/pulse/brexit-betrayed-how-ai-social-media-hijacked-democracy-dr-khalid-al--8fwge.

74. Vasist, P. N., Chatterjee, D., and Krishnan, S. "The Polarizing Impact of Political Disinformation and Hate Speech: A Cross-country Configural Narrative." *Information Systems Frontiers* (2023): 1–26. https://doi.org/10.1007/s10796-023-10390-w.

75. Barrett, P., Hendrix, J., and Sims, G. "How Tech Platforms Fuel U.S. Political Polarization and What Government Can Do About It." *Brookings*, September 27, 2021. https://www.brookings.edu/articles/how-tech-platforms-fuel-u-s-political-polarization-and-what-government-can-do-about-it/.

76. Cinelli, M., Morales, G. D. F., Galeazzi, A., Quattrociocchi, W., and Starnini, M. "The Echo Chamber Effect on Social Media." *Proceedings of the National Academy of Sciences* 118, no. 9 (2021). https://doi.org/10.1073/pnas.2023301118.

77. Tversky, A., and Kahneman, D. "Judgment under Uncertainty: Heuristics and Biases." *Science* 185, no. 4157 (1974): 1124–31. https://www2.psych.ubc.ca/~schaller/Psyc590Readings/TverskyKahneman1974.pdf.

78. Department of Homeland Security. "Increasing Threat of Deepfake Identities." *DHS*, 2023. https://www.dhs.gov/sites/default/files/publications/increasing_threats_of_deepfake_identities_0.pdf.

79. Freelon, D., and Wells, C. "Disinformation as Political Communication." *Political Communication* 37, no. 2 (2020): 145–56. https://doi.org/10.1080/10584609.2020.1723755.

80. Griskevicius, Vladas, and Douglas T. Kenrick. "Fundamental Motives: How Evolutionary Needs Influence Consumer Behavior." *Psychological Review* 119, no. 2 (2012): 376–401. https://doi.org/10.1037/gpr0000056.

81. Adisa, D. "Everything You Need to Know about Social Media Algorithms." *Sprout Social*, October 30, 2023. https://sproutsocial.com/insights/social-media-algorithms/.

82. Azzimonti, M., and Fernandes, M. "Social Media Networks, Fake News, and Polarization." *European Journal of Political Economy* 76 (2022): 102256. https://doi.org/10.1016/j.ejpoleco.2022.102256.

83. Roscini, F. "How the American Media Landscape Is Polarizing the Country." *The Pardee Atlas Journal of Global Affairs*, 2022. https://sites.bu.edu/pardeeatlas/advancing-human-progress-initiative/back2school/how-the-american-media-landscape-is-polarizing-the-country/.

84. Singer, P. W., and Emerson T. Brooking. "Weaponized Narrative Is the New Battlespace." *Defense One*, January 3, 2017. https://www.defenseone.com/ideas/2017/01/weaponized-narrative-new-battlespace/134284/.

85. Helbing, Dirk, Bruno S. Frey, Gerd Gigerenzer, Ernst Hafen, Jeroen van den Hoven, Roberto V. Zicari, and Andrej Zwitter. "Will Democracy Survive Big Data and Artificial Intelligence?" *Scientific American*, February 25, 2017. https://www.scientificamerican.com/article/will-democracy-survive-big-data-and-artificial-intelligence/.

Chapter 7

86. The Jed Foundation. "Mental Health and Suicide Statistics." *The Jed Foundation*, February 18, 2022. https://jedfoundation.org/mental-health-and-suicide-statistics/.

87.   World Health Organization. "'Depression: Let's Talk,' Says WHO, as Depression Tops List of Causes of Ill Health." *World Health Organization*, March 30, 2017. https://www.who.int/news/item/30-03-2017--depression-let-s-talk-says-who-as-depression-tops-list-of-causes-of-ill-health.

88.   Hollingworth, W., Fawsitt, C. G., Dixon, P., Duffy, L., Araya, R., Peters, T. J., Thom, H., Welton, N. J., Wiles, N., and Lewis, G. "Cost-Effectiveness of Sertraline in Primary Care According to Initial Severity and Duration of Depressive Symptoms: Findings from the PANDA RCT." *PharmacoEconomics - Open* 4, no. 3 (2019): 427–38. https://doi.org/10.1007/s41669-019-00188-5.

89.   National Alliance on Mental Illness. "Mental Health by the Numbers." *National Alliance on Mental Illness*. Accessed September 10, 2024. https://www.nami.org/about-mental-illness/mental-health-by-the-numbers/.

90.   Rodriguez, Jonathan A., Daniel Lipsitz, and Eleni Linos. "Telemedicine for Diabetic Foot and Pressure Ulcers: A Systematic Review." *Journal of Telemedicine and Telecare* 27, no. 9 (2021): 515-524. https://www.ncbi.nlm.nih.gov/pmc/articles/PMC8283615/.

91.   Logoffmovement. "How the Instinctual Fulfillment of Social Media Algorithms Has Led to a Loss of Self-Control." *LOG OFF*, June 12, 2023. https://www.logoffmovement.org/2023/06/12/how-the-instinctual-fulfillment-of-social-media-algorithms-have-led-to-a-loss-of-self-control/.

92.   "Is Your Phone Secretly Hearing All Your Conversations? Here's What to Do." *The Times of India*, February 23, 2024. https://timesofindia.indiatimes.com/etimes/trending/is-your-phone-secretly-hearing-all-your-conversations-heres-what-to-do/photostory/107940821.cms.

93.   Ashleylatimer. "3 Signs That You May Be Codependent with Social Media." *Elevated With Ashley*, January 25, 2022. https://elevatedwith-ashley.com/codependent-with-social-media/.

94.   Harvard Second Generation Study. "Harvard Second Generation Study." *Harvard Study*. Accessed June 2024. https://www.adultdevelopmentstudy.org/.

95.   Bowman, Alisa. "Social Media's Effects on the Teen Brain." *Mayo Clinic Press*, September 5, 2023. https://mcpress.mayoclinic.org/parenting/social-media-affects-teens-brains/.

96.   Schonfeld, A., McNiel, D., Toyoshima, T., and Binder, R. "Cyber-bullying and Adolescent Suicide." *Journal of the American Academy of Psychiatry and the Law* 51, no. 1 (2023): 112–19. https://doi.org/10.29158/JAAPL.220078-22.

97.   Mosley, T., and Hagan, A. "'The Social Dilemma' Director Says the Internet Is Undermining Democracy." *WBUR*, September 18, 2020. https://www.wbur.org/hereandnow/2020/09/18/social-dilemma-director.

Chapter 8

98.   Hilliard, J. "Social Media Addiction." *Addiction Center*, October 26, 2023. https://www.addictioncenter.com/drugs/social-media-addiction/.

99.   Haynes, T. "Dopamine, Smartphones & You: A Battle for Your Time." *Science in the News* (Harvard University), May 1, 2018. https://sitn.hms.harvard.edu/flash/2018/dopamine-smartphones-battle-time/.

100.   Centers for Disease Control and Prevention. "Drug Overdose Deaths." *CDC National Center for Health Statistics*. Last updated August 2023. https://www.cdc.gov/nchs/nvss/vsrr/drug-overdose-data.htm.

101.   Substance Abuse and Mental Health Services Administration. *2021 National Survey on Drug Use and Health: Detailed Tables*. Updated January 3, 2023. https://www.samhsa.gov/data/sites/default/files/reports/rpt39443/2021NSDUHFFRRev010323.pdf.

102.   World Health Organization. *Global Status Report on Alcohol and Health and Treatment of Substance Use Disorders*. June 25, 2024. https://www.who.int/news/item/25-06-2024-global-status-report-on-alcohol-and-health-and-treatment-of-substance-use-disorders.

103.   Miller, J. "Social Media Addiction Statistics (2023) - Who Is Most at Risk?" *Addiction Help*, August 15, 2023. https://www.addictionhelp.com/social-media-addiction/statistics/.

104.   Cleveland Clinic. "Addiction." *Cleveland Clinic*, March 16, 2023. https://my.clevelandclinic.org/health/diseases/6407-addiction.

105.   "The Social Dilemma: Social Media and Your Mental Health." *McLean Hospital*, March 29, 2024. https://www.mcleanhospital.org/essential/it-or-not-social-medias-affecting-your-mental-health.

106.   "Is Our Aversion to Pain Killing Us?" *Psychology Today*. Accessed July 14, 2024. https://www.psychologytoday.com/us/blog/compassion-matters/201203/is-our-aversion-pain-killing-us.

107. Waters, J. "Constant Craving: How Digital Media Turned Us All into Dopamine Addicts." *The Guardian*, August 22, 2021. https://www.theguardian.com/global/2021/aug/22/how-digital-media-turned-us-all-into-dopamine-addicts-and-what-we-can-do-to-break-the-cycle.

108. Miller, S. "The Addictiveness of Social Media: How Teens Get Hooked." *Jefferson Health*, June 2, 2022. https://www.jeffersonhealth.org/your-health/living-well/the-addictiveness-of-social-media-how-teens-get-hooked.

109. "China: The Rise of Digital Repression in the Indo-Pacific." *Article 19*, April 18, 2024. https://www.article19.org/resources/china-the-rise-of-digital-repression-in-the-indo-pacific/.

110. "Social Media Addiction Statistics." *The Lanier Law Firm*. Accessed July 2024. https://www.lanierlawfirm.com/social-media-addiction/statistics/.

Chapter 9

111. ScienceDirect. "Actively Open-Minded Thinking." *ScienceDirect Topics*. Accessed September 10, 2024. https://www.sciencedirect.com/topics/psychology/actively-open-minded-thinking.

112. Kahneman, Daniel. *Thinking, Fast and Slow*. New York: Farrar, Straus and Giroux, 2011.

Chapter 10

113. Borealis AI. "Quantum AI: The Next Revolution in Computing." *Borealis AI*, August 3, 2021. https://borealisai.com/research/quantum-ai-revolution.

114. IBM. "IBM Unveils 400 Qubit-Plus Quantum Processor and Next-Generation IBM Quantum System Two." *IBM Newsroom*, November 9, 2022. https://newsroom.ibm.com/2022-11-09-IBM-Unveils-400-Qubit-Plus-Quantum-Processor-and-Next-Generation-IBM-Quantum-System-Two.

115. Butler, N., Quigg, Z., Bates, R. et al. The Contributing Role of Family, School, and Peer Supportive Relationships in Protecting the Mental Wellbeing of Children and Adolescents. *School Mental Health* 14, 776–788 (2022). https://doi.org/10.1007/s12310-022-09502-9

116. McKinsey Global Institute. (2017). *How artificial intelligence can deliver real value to companies.* McKinsey & Company. Retrieved from https://www.mckinsey.com/capabilities/quantumblack/our-insights/how-artificial-intelligence-can-deliver-real-value-to-companies

117. Simply Psychology. (2023). *Gardner's Theory of Multiple Intelligences.* Retrieved from https://www.simplypsychology.org/multiple-intelligences.html

118. King, Martin Luther, Jr. *"Remaining Awake Through a Great Revolution."* Speech, National Cathedral, Washington, D.C., March 31, 1968. https://www.nps.gov/mlkm/index.htm.

Image Attribution Statement

All images in this book were generated using DALL·E, an AI model created by OpenAI. The creative direction and concept development behind these images are original to the work presented in *humAIn*. All rights and ownership of the images, as part of this book, are fully retained and protected under the copyright of *humAIn*.

INDEX

Milton Keynes UK
Ingram Content Group UK Ltd.
UKHW040403111224
452348UK00004B/392